RhomboSTEAM
K-2 adventures in geometry

for

3D Shapes

SECOND GRADE

STEAMBRIDGE PRESS

Charlotte North Carolina
Portland Oregon

Copyright 2014 by Jeannie Ruiz and Elizabeth Spivey Stripling

Published by STEAMBridge Press
2901 North Davidson Street
Charlotte, NC 28201

All rights reserved. No part of this publication may be reproduced, stored in a retrieval system, or transmitted in any form for any reason, recording or otherwise, without the prior written permission of STEAMBridge Press.

ISBN 978-1-942357-28-5

Photo permissions and authors, illustrators, designers, photographers available for download at the URL.

http://www.streambridgepress.com

STEMWORKS EXEMPLARY PROGRAM STATUS

Ten80 Education and its programs are included in the STEMWorks Database by Change the Equation, a non-partisan, CEO-led initiative to connect and align efforts to improve STEM learning. Ten80 Education programs met the highest standards for excellence and exemplary STEM project based programming through rigorous examination by WestEd, an independent nonprofit research organization.

RECOMMENDED BY REAL TEACHERS

We are thoroughly enjoying our STEAM lessons. The program is excellent! My students have loved making the Rhombis and getting to build with shapes and tape. It has been a big hit! Thanks for everything!
Alicia Stenard
Mater Christi in Albany School District

"We need more projects like this that carry on for a longer number of weeks. My teachers used scope and sequence suggestions, and the program ran after school for a semester. We could have worked on this a lot longer."
Karen James
Druid Hills Elementary in Charlotte, NC

"I'm excited for our students to get their hands on real STEAM projects. There isn't much out there for Kindergarten and First Grade."
Sandy Mettler
Fort Worth Schools in Dallas, TX

TO THE EDUCATORS USING RHOMBOSTEAM,

First of all, thank you for taking the leap. You may have completed the 2D Modules, and you are knee-deep in project-based learning. You know that it requires a commitment on the part of the teacher. It can be messy. It can mean more planning and more time spent gathering odds and ends that make a project work. The time and energy you spend developing the imaginative responses and ability to problem solve will pay off tenfold. Please pass this information on to new teachers. You can also reassure the veterans that this is not a fly-by-night movement, and the shift is permanent. It has to be if our students are to compete on the international stage.

There is significant evidence in our own research that indicates a need for memorization and rote learning IN ADDITION TO LONG-RANGE PROJECTS and innovative problem-solving. A decent project uses the challenge as a springboard for planning skills-based lessons. A great project urges kids to seek skills and knowledge that you can offer because they want to succeed in solving the problem you've posed. Quick recall of facts is necessary for innovative thinking. Kids don't need to stop and look up simple facts or count on teddy bears when we want them to focus on collecting data and making reasonable decisions.

Projects need outcomes. They need number-driven goals that qualify success in the challenges. Let kids get creative and try things that will fail. Did I say to let them fail? Let them fail. Innovators fail all the time. Some of our greatest inventions are the result of absolute and profound failures. Learning to use the failure as a tool rather than an emotional trial is where they are headed. It takes a few falls before learning to ride a bike. They will make a few bouncy balls on the road to a great slimy wall-walker.

The question arises as to why we chose to call students "designers" rather than "engineers" or "scientists." Designers incorporate a wide variety of communication skills as well as industry expertise. Because communication of ideas among students is integral to the success of any project-based curriculum, the students become Rhombi's designers.

Ten80 celebrates the amazing lessons we learn from you. When you develop a new technique or unique approach to the material, let us hear from you. The resource site grows stronger as teachers share their ideas with each other.

Sincerely,

Jeannie Ruiz
President of Ten80 Elementary

DESIGNERS, WILL YOU HELP ME?

Designers collect, analyze, and synthesize data guided by specifications.

Designers make clear and concise recommendations using drawings, verbal descriptions, and math models.

The Industrial Designers Society of America's definition of industrial design describes design as "the professional service of creating and developing concepts and specifications that optimize the function, value and appearance of products and systems for the mutual benefit of both user and manufacturer."

Rhombi asks her friends to offer assistance in design. Students collect, analyze and make data-based decisions. They must share their concepts verbally, using drawings, and through models (scale and math).

RHOMBI'S ADVENTURES IN 3D
CONTENTS

ACTIVITY	HANDOUT	READ ALOUD	
RHOMBI'S HOUSE *PAGE 18*			
WEATHER	23	140	WIND ON THE WINDOWS	60
DESIGN	24	141	ACCIDENTAL INVESTIGATIONS	62
STRUCTURES	25	142	THE TINY HOUSE	64
CUBISM	26	143	CUBE WORLD	66
PERSPECTIVE	27	144	DO YOU KNOW THE QUADRILATERALS	68
UP ON THE ROOF *PAGE 42*			
WATER CYCLE	47	145	WATER IS AS WATER DOES	72
PYRAMIDS	48	146	CASE OF THE DISAPPEARING SAND	74
ROOF HEIGHT	49	147	CIRCLE, TRIANGLE, SQUARE	76
MIXTURES	50	148	I LOVE HOMES	78
EROSION	51	149	THE CAMEL AND THE PYRAMID	80
NEW HOME FOR PET *PAGE 64*			
POLYGONS	67	146	RICKETY RACKETY	34
PRESENTATIONS	68	147	CITY OF THREE ANGLES	36
GROWING	69	148	TRAVELING TRIANGLE	38
TRIANGLES IN ART	70	149	FOX'S FOREST	40
WHAT'S YOUR ANGLE	71	150	TRI, TRI AGAIN	42
PET'S CELEBRATION *PAGE 87*			
MAGNIFY THE PENTAGNO	91	151	HUNTING FOR PENTAGONS	44
TOOLS THROUGH TIME	92	152	TOOL TIME	46
MATERIALS	93	153	CREATIVE DESIGN	48
STORY IN 5 BOXES	94	154	DESIGN TME	50
WALK THE PENTAGON	95	155	PENTAGON, USA	52
RHOMBI'S GALLERY *PAGE 105*			

INTRO TO STEAM

6	LONG RANGE PROJECT
7	MATH MODELING IN K-2
8	PROJECT BASED LEARNING
9	GREAT S.T.E.A.M. TEACHER CHECKLIST
10	CONFIDENT S.T.E.A.M. STUDENTS
11	TEACHER MISCONCEPTOINS
13	DEFINING S.T.E.A.M.
14	BUILDING S.T.E.A.M. IN CLASSROOMS
16	GEOMETRY IS THE THREAD
17	USING THE MATERIALS

APPENDIX

127	NETS
129	POLYGON TRACING PAGES
138	POLYGON BACKGROUND
139	STEMVESTIGATIONS HANDOUTS
165	MINDBUGS IN MEASUREMENT
176	NGSS RUBRICS
177	PRE / POST ASSESSMENTS
186	WORKS CITED
187	PLANNING PAGES
189	COMMON CORE CURRICULUM K-2
197	SCOPE AND SEQUENCE

RHOMBI'S ADVENTURES IN 3D
UNITS OVERVIEW

RHOMBI'S HOUSE (Play with "SOCK IT TO ME.")   18

Designers will help Rhombi build a new cube home.

SCIENCE: Build an anemometer to check wind speed and direction.
    23

TECHNOLOGY: Create a wind-proof barrier.
    24

ENGINEERING: Fold a cube from its net. Design and customize personal nets.
    25

ART: Create a new picture with shapes.
    26

MATHEMATICS: Build and draw cubes in perspective studies.
    27

UP ON THE ROOF (Play with "LEARN TO JUGGLE.")   35

Designers will help Rhombi build a pyramid roof for her house.

SCIENCE: Test various surfaces and materials for water resistance.
    47

TECHNOLOGY: Visit a real pyramid in the virtual world.
    48

ENGINEERING: Build the tallest pyramid possible.
    49

ART: Make sturdy clay bricks.
    50

MATHEMATICS: Build a pyramid using clay bricks, and experiment with the elements.
    51

RHOMBI'S ADVENTURES IN 3D
UNITS OVERVIEW

NEW HOME FOR PET (Play with "HOOP AND HOLLAR.")   69

Designers will help Rhombi create a room for pet.   70

SCIENCE: Build and design habitats for pets.   71

TECHNOLOGY: students create their "wild selves" with the NY Zoo website.   72

ENGINEERING: Investigate structures with platforms. Build square and triangle based creations.   73

ART: Create your own tesselation patterns.   74

MATHEMATICS: Create a size and scale chart.   75

PET'S CELEBRATION (Play with "HOPSCOTCH.")   87

Designers will help Rhombi create a space for Pet's celebration.

SCIENCE: Make creamy dreamy pet-safe treats that are also yummy for humans.   91

TECHNOLOGY: Use a digital scale to weigh various classroom objects.   92

ENGINEERING: Build and test the strength of columns.   93

ART: Create beautiful polygon mosaics.   94

MATHEMATICS: Create a yardstick / length string art.   95

RHOMBI'S GALLERY   105

Designers will help Rhombi build her business.

RHOMBI'S ADVENTURES IN 3D
LONG RANGE PROJECT

If we had a floating house (school, city, state) that could go anywhere on the ocean, where would we want to be?

Instead of being planted in one place, what if we could take our house (school, city, state) with us? Turtles do. No matter where the turtle goes, it's always home.

Will we stay in the same place all the time, or move with currents, weather patterns and seasons?

Build a room, house, school, city, state, or country as a floating island.
Basic elements are squares and triangles which combine (add) to form a solid. Students begin to seek skill in STEAM subjects as a way to solve problems and overcome challenges... as long as the project is interesting enough to keep their attention.

Begin investigating:
- How much milk is delivered to your school everyday?
- Where does it come from?
- What is the source of milk on the floating city-state island?
- How many people can live in a defined habitat? What do they require for survival and comfort?
- How will you move from island to land?
- What happens to structures in various kinds of weather?
- Can we stay out of the way of huge storms over the ocean?
- Do we want to move with the seasons?
- Will it move? How fast do things move?
- If we move, what will fuel the island?
- How much will all that cost, by the way?
- Where is the funding source?

and on... and on... and on...

Since we can't build full-size during the planning phase, we'd better start with scale models. The models will save time, energy, materials, cost and headache. Let's start with the basic information on geometry, polygons and structures.

Introducing RhomboSTEAM.
Once upon a time, there was Rhombi. She loved shapes..."

RHOMBI'S ADVENTURES IN 3D
MATH MODELING IN K-2 SETTINGS

DEFINING THE MATH MODEL
Mathematics can be used to "model", or represent, how the real world works. A model is not the real thing, but it should be accurate enough to be useful. Math models connect abstract numbers to real experience. Math models save time, supplies, and aggravation. For example, Rhombi needs to build a full-scale cart. Since the class cannot build the full sized vehicle, it will develop a small scale version and extrapolate from there. They use their background knowledge to begin the math model.

MOVE FROM EXPERIENCE TO SYMBOLS
The study of science and math started in an effort to understand actions that could be observed directly. In order to make science and math a useful performing art, instruction must start with real experience and gradually build toward an abstract representation. Doing things in the reverse chronological order produces nearly useless results.

KNOW WHERE THE LESSON IS HEADED
Education separates experience from the learning process. We could learn a few things from our own history. Math development progressed along its slow, continuous track: caveman to calculus. Here, in the late 20th century, we expect second graders to operate at abstract levels using a symbolic language that took mathematicians thousands of years to develop. Educators are asked to start with the algorithm –the abstraction- and work backwards to the concrete. We wonder why math is such a dreaded subject. Neither the teacher nor the student knows where the symbols are supposed to take them.

STEAM USES NATURAL EXTENSIONS
Abstract models of STEAM phenomena should grow as a natural extension of what is felt by direct experience. The abstraction must be an immediate and obvious representation of what students already know at every level. Drive a radio controlled car. Build the shapes as you memorize them. Investigate, collect data and analyze results. From reality to abstraction, STEAM learning needs more teachers who understand that doing more means going further, faster.

DEFINE YOUR MATHEMATICAL MODEL
Identify the problem, define the terms in your problem, & draw diagrams where appropriate. Begin with a simple model, stating assumptions that you make as you focus on particular aspects of the phenomenon. Identify important variables and constants. Determine how they relate to each other. Develop the equation(s) that express relationships between variables & constants.

RHOMBI'S ADVENTURES IN 3D
PROJECT BASED LEARNING

Learning shapes is great. Doing short-term interactive projects that involve shapes is even better. Turning those projects into inquiry-based challenges takes things a step further. Involving students in long-range projects that require multiple steps, depth of knowledge, and true integration of science and math is the goal. The National Science Foundation funded Innovative Technology Experiences for Teachers (ITEST) research that developed a list of criteria for excellence in project based learning.

CRITERIA FOR EXCELLENCE IN PROJECT BASED LEARNING

- Focus on activities that motivate students to develop interest and skills in STEM careers.

- Provide intensive engineering experience of a long duration (academic year and/or summer proportional to the nature of the program).

- Use technlogy tools and hands-on learning.

- Involve industry mentors, scientists, and engineers.

- Incorporate students' self-regulated learning within specific engineering design experiences.

- Offer innovative, high interest STEM activities.

- Involve participants in the designing and building of artifacts.

- Use scientific integration and/or the engineering design process.

- Engage students in project-based and/or problem-based learning.

- Develop participants' technical skills (as appropriate to the project).

- Make explicit connections between (STEM) academics and project.

- Build in success experiences and opportunities to learn from failure.

- Explore technical careers from technicians to engineers and engineering management.

RHOMBI'S ADVENTURES IN 3D
GREAT STEAM TEACHER CHECKLIST

1. **DO MORE THAN IMPART KNOWLEDGE** Children will teach themselves if offered the opportunities to explore and problem solve. Teach by posing questions. Teach by developing curiosity. Teach by guiding learners with interactive experiences. Teach to learn. Learn as much as you teach. Adults often take so much control of the teaching process that children resort to mimicry as a way of pleasing their mentors. Every child learns at a unique pace and with a unique set of circumstance; therefore each child requires a unique path. If educators view learning as a problem-solving process, learning becomes an individual journey.

2. **DISLODGE THE MISCONCEPTIONS** Every child enters school with a unique set of skills and preconceptions about learning. Every child has already formed opinions based on observation, mimicry, and the outcome of challenges (how were successes and failures handled by adults). Some facts that lodge in a child's mind are erroneous. Other ideas are keen observations of the natural and technological world. Facilitate learning. Increase knowledge. Guide inquiry. Open minds to the possibilities. Offer multiple opportunities to learn and approaches to problem solving as a way of addressing misconceptions.

3. **USE WHOLE GROUP SPARINGLY** There are rare teachers with the ability to engage learners as a group. Generally, this interaction has a fast-paced Q&A quality and involves a points system or prizes. The up-side is that learners are involved as a group and learn courteous behavior. The down-side is that educators are rarely able to get individual feedback or offer reinforcement / stimulation. Even technologically based feedback systems are an impersonal way to discover a student's strong and weak areas.

4. **LEARN MORE WITH SMALL GROUPS** Teachers that interact in small groups with students on a regular basis know far in advance of testing exactly what the outcomes will be. Anxiety lessens for teachers and learners. Differentiation becomes possible, and students no longer "tune out" the instructor when the instructor becomes a facilitator. It's hard to ignore the teacher when she's sitting at your table asking questions and challenging assumptions. It's difficult to avoid detection if the investigation requires active participation. For the learner, there is no more "tuning out" or sleep-walking through the day.

5. **ASSESS STUDENTS REGULARLY** There are no surprises when learners take part in authentic and active assessment with reinforcement for some and mental stimulation for others. Great teachers know well ahead of time how students will perform on standardized tests.

RHOMBI'S ADVENTURES IN 3D
CONFIDENT STEAM STUDENTS

Children View Themselves as Scientists in the Process of Learning.
1. They look forward to doing science.
2. They demonstrate a desire to learn more.
3. They seek to collaborate and work cooperatively with their peers.
4. They are confident in doing science;
   they demonstrate a willingness to modify ideas, take risks, and display healthy skepticism.

Children Accept an "Invitation to Learn" & Readily Engage in The Exploration Process.
1. Children exhibit curiosity and ponder observations.
2. They move around selecting and using the materials they need.
3. They take the opportunity and the time to "try out" their own ideas.

Children Plan and Carry Out Investigations.
1. Children design a way to try out their ideas, not expecting to be told what to do.
2. They plan ways to verify, extend or discard ideas.
3. They carry out investigations by: handling materials, observing, measuring, and recording data.

Children Communicate Using a Variety of Methods.
1. Children express ideas in a variety of ways: journals, reporting out, drawing, graphing, charting, etc.
2. They listen, speak and write about science with parents, teachers and peers.
3. They use the language of the processes of science.
4. They communicate their level of understanding of concepts that they have developed to date.
   Children Propose Explanations and Solutions and Build a Store of Concepts.

Children offer explanations from a "store" of previous knowledge.
(Alt Frameworks, Gut Dynamics).
1. They use investigations to satisfy their own questions.
2. They sort out information and decide what is important.
3. They are willing to revise explanations as they gain new knowledge.

Children Raise Questions
1. Children ask questions (verbally or through actions).
2. They use questions to lead them to investigations that generate further questions or ideas.
3. Children value and enjoy asking questions as an important part of science.

Children Use Observation.
1. Children observe, as opposed to just looking.
2. They see details, they detect sequences and events; they notice change, similarities and differences, etc.
3. They make connections to previously held ideas.

Children Critique Their Science Practices.
1. They use indicators to assess their own work
2. They report their strengths and weaknesses.
3. They reflect with their peers.

"Inquiry Based Science: What Does It Look Like?" Connect Magazine (published by Synergy Learning), March-April 1995, p. 13.

RHOMBI'S ADVENTURES IN 3D
TEACHER MISCONCEPTIONS

These misconceptions were identified based on in-depth interviews with early childhood teachers about the key issues in early mathematics education as well as researchers' experiences in teaching early childhood students, conducting workshops with early childhood teachers (Ginsburg, Jang, Preston, VanEsselstyn & Appel, 2004; Ginsburg et al., 2006), working with them in early childhood classrooms, and engaging in informal conversations with them. The study also based descriptions of the myths on available research literature (Ginsburg, Lee & Boyd, 2008). The nine misconceptions defined by Joon Sun Lee of Hunter College, The City University of New York and Herbert P. Ginsburg of Teachers College, Columbia University are:

1. Young children are not ready for mathematics education.
2. Mathematics is for some bright kids with mathematics genes.
3. Simple numbers and shapes are enough.
4. Language and literacy are more important than mathematics.
5. Teachers should provide an enriched physical environment, step back, and let the children play.
6. Mathematics should not be taught as stand-alone subject matter.
7. Assessment in mathematics is irrelevant when it comes to young children.
8. Children learn mathematics only by interacting with concrete objects.
9. Computers are inappropriate for the teaching and learning of mathematics.

These misconceptions often interfere with understanding and interpreting the new recommendations on sound early childhood mathematics education, and become subtle (and sometimes overt) obstacles to implementing the new practices in the classrooms (Richardson, 1996).

A WORD FROM TEN80 TO TEACHERS
Become aware of your personal biases. Just because you had poor math instruction does not mean math is hard.

The language of math cannot be taught with the same symbolic approach as the language of reading and writing. Because everyone talks and practices the art of communication, teachers assume that words are the only way to demonstrate proficiency in every field including math and science. Math and science are performing arts that must be demonstrated in performance. Math and science must use pictures and numbers depicting an actual experiment that is a test of the student's abstract model of some aspect of reality. In this way, students build their own understanding of the patterns that underlie math and science. (Ten80 2002)

RHOMBI'S ADVENTURES IN 3D
DEFINING STEAM

STEAM (science, technology, engineering, arts, mathematics) is just now becoming an accepted acronym among teachers and is seen as a somewhat mysterious –almost magical- fix for all the problems American schools and businesses are facing. Perhaps we should start with what STEAM is not:
- STEAM is not a conglomeration of lessons culled from five subject areas.
- STEAM is not something you can cover in a unit or two every year.
- STEAM is not a 5-day or 10-day curriculum piece.
- STEAM cannot be delivered in the traditional pencil-and-paper method by a sage on the stage standing behind the classroom lectern or seated behind a grand old wooden desk. There are no "Buehler..? Buehler..?" moments in a STEAM environment. *(reference the movie Ferris Buehler's Day Out)*

SO WHAT IS STEAM?
STEAM is a philosophy of teaching, learning and working. STEAM is more than just the sum of the words in its acronym.

In other words, do real stuff to find real answers to real problems.

STEAM IS PRETTY SIMPLE TO STEAM PROFESSIONALS
What has become a raging debate among educators seems pretty simple to the professionals that actually work in STEAM fields. Academicians will spout research from their PhD thesis and published articles in journals, etc. Ask a scientist, a mathematician, an engineer or a programmer how to prepare kids for a STEAM workplace and they'll tell you to make kids DO something.

SET UP CLASSROOMS FOR DISCOVERY
While activity centers are typical in kindergarten settings, they have all but disappeared by second grade. Bring them back. Small group interaction is optimal for RhomboSTEAM as it is for any STEAM program. Why do you want to use stations or centers in the classroom?
- Simplifies planning - plan over a week rather than day by day...
- Offers opportunities for more cohesive lessons...
- Allows teachers to hold small group sessions daily...
- Encourages teamwork...
- Simplifies classroom management... *
- STEAM increases the ability to draw connections among various concepts.

* reference   Harry Wong and **The Effective Teacher**
https://www.effectiveteaching.com

RHOMBI'S ADVENTURES IN 3D
BUILDING STEAM CLASSROOMS

According to corestandards.org, "The Common Core State Standards provide a consistent, clear understanding of what students are expected to learn, so teachers and parents know what they need to do to help them. The standards are designed to be robust and relevant to the real world, reflecting the knowledge and skills that our young people need for success in college and careers. With American students fully prepared for the future, our communities will be best positioned to compete successfully in the global economy."

Support Common Core Standards and Learning Goals

STEAM PROJECT STATION
Support the ongoing STEAM project with a permanent station and construction materials. Recycled and upcycled materials should be available for use along with tools, paper, pencil and data collection opportunities.

ART, MUSIC, DRAMA
Encourage discussion of STEAM subjects through the arts as well as investigations involving art inventions, art history, genres, musical instruments and sound, etc. Create a "stage" where students give skits, give presentations, teach lessons and make up songs about classroom learning.

BLOCK PLAY
Keep blocks available all the time along with small scale toys at different scales. Add books about world architecture, habitats, and building projects.

NONFICTION LIBRARY
Facilitate the transition from fiction and storybooks to gleaning facts from informational text. Fill your library with fiction AND non-fiction. Encourage students to write informational text and to keep journals for science and math.

STEAM CHARACTER ROLE PLAY
Most kindergarten classrooms include an area for dress-up and pretend. How many first and second graders would still benefit from this type of role-play? Shoot for higher level characterization with lab coats, construction hats, medical scrubs, artist's smocks, etc.

RHOMBI'S ADVENTURES IN 3D
GEOMETRY IS THE THREAD

The word "geometry" comes from the Greek words for "Earth" and "measure." Geometry was first used to measure and chart the length, area and shape of land surfaces.

GEOMETRY is the single math thread that runs through Common Core from Kindergarten to 12th grade. RhomboSTEAM promotes STEAM and the Art of STEM through an authentic experience in engineering problem solving built on a platform of geometric constructions, analytical geometry, and engineering design. Math is a gateway to success in professional fields: speaking the language of mathematics fluently means that students may define their career paths by choice, not resignation.

~   Jeannie Ruiz

RhomboSTEAM Developer and Director of STEM Ten80 Foundation

THE MATH GENE
Contrary to popular belief, no one is born with a genetic predisposition toward math. While there is no "math gene" in humans (Keith Devlin, The Math Gene) mathematics is a learned skill for everyone just like basketball, playing the French Horn, or ballet. Humans have brains which are already wired to enable math acquisition (Stanislas Dehaene, Number Sense 1997). Infants, chimps, dogs, and birds can recognize groups of objects up to three and decide that an object has been taken away or added. That is not "doing math," but it is a qualitative foundation on which to build understanding. Math instruction in school develops the initial qualitative sense of space, time, and quantity into the descriptive -and highly useful- language of mathematics. Children should learn all things in context.

MULTI YEAR FLUIDITY
Geometry as a foundation allows the educator to implement multi-year projects that students enhance year after year. The essence of the project is still fluid, while monthly, weekly, and daily objectives are met for the grade level.

NEXTGEN SCIENCE STANDARDS - STRUCTURE AND FUNCTION
Engineering and Design: Students should understand that "the shape and stability of structures of natural and designed objects are related to their function(s). (K-2-ETS1-2)."

RHOMBI'S ADVENTURES IN 3D
EACH STEMVESTIGATION IS A PROJECT OR ACTIVITY

Approach each STEMvestigation as a project OR as an activity. Projects avoid the "cookie-cutter" recipe process and encourage decision-making based on data collected by the student. Activities give instructions step-by-step. Choose based on time available and depth give to the subject.

READ Engage students with literature. Each activity offers a read-aloud option or video / audio jumpstart.

DISCUSS The MindBug is a misconception in student thinking; procedural understanding that worked in one situation but is incorrectly applied to other skills. Like a computer virus, MindBugs spread through a student's system. Discussion offers a chance for the teacher to discover mindbugs hidden in student ideas.

Example In Science:
Children often think that air weighs nothing because they see scales measuring zero from an early age.

Example in Math:
Students line the numbers up on the right to add. This works until 5th grade when they attempt to add decimals.

EXPLAIN Each activity includes a brief background for the purpose of introducing the subject matter.

MOVE Get students up and moving to activate that extremely useful muscle memory.

INVESTIGATE

Although each STEMvestigation stands alone, together they also build knowledge required to meet the long-range project. In this Module, students work through basic understanding and vocabulary as well as Common Core Curriculum math and english skills. Each STEMvestigation is a project. All projects are used in the Unit Challenge.

ELA Build skills in communication, writing and speaking.

RHOMBI'S HOUSE

RHOMBI'S ADVENTURES IN 3D
UNIT 1: CUBES

RHOMBI'S ADVENTURES IN 3D
UNIT 1: RHOMBI'S HOUSE

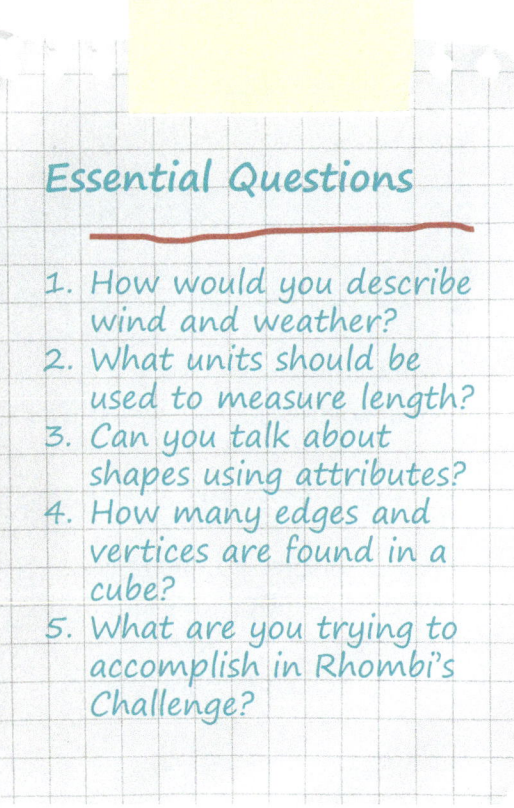

Essential Questions

1. How would you describe wind and weather?
2. What units should be used to measure length?
3. Can you talk about shapes using attributes?
4. How many edges and vertices are found in a cube?
5. What are you trying to accomplish in Rhombi's Challenge?

ENDURING UNDERSTANDING

Two dimensional shapes create new shapes.

"Halves" means two equal parts."

Circles and rectangles break into equal parts.

Breaking circles or rectangles into more equal parts means that the parts will be smaller.

Three objects may be put in order from longest to shortest by comparing their lengths.

The length of an object may be discussed using whole numbers.

A smaller object may be used as a measurement tool when measuring longer lengths.

Larger numbers consist of tens and ones.

Organizing information helps in discussion.

RHOMBI'S HOUSE

Focus on squares and cubes. Construct an anemometer and create a windy record. Investigate mistakes that worked. Develop a solid wall seam using adhesives and materials. Take a closer look at geometric shapes in real life, and gain a better perspective in math.

Students will use basic measurement, understanding of number, place value, data collection, and data application in problem solving. Two-dimensional shapes will become three-dimensional shapes with a purpose.

Rhombi's house on the hill is pretty rickety. She needs assistance in designing and building a new, strong, wind-resistant home. Students will collect data, design and construct a scale model that withstands the hair-dryer test.

cube
box shaped solid object formed with six squares

forecast
to calculate or predict the future

weather
The state of the atmosphere at a given time and place, with respect to variables such as temperature, moisture, wind velocity, and barometric pressure. (Free Dictionary)

RHOMBI'S HOUSE
CURRICULUM CONNECTIONS

Rhombi needs a new cube home to withstand windy weather.

ART CONNECTIONS - Cubism

Cubism was created by Pablo Picasso and Georges Braque in Paris between 1907 and 1914. The French art critic Louis Vauxcelles coined the term Cubism. Vauxcelles called the geometric forms in Braque's abstracted works "cubes." Cubist works emphasize the flat two-dimensional nature of canvas. Cubists reduced and fractured objects into geometric forms. Sculptors included Alexander Archipenko and Raymond Duchamp-Villon.
source: http://www.metmuseum.org

TECHNOLOGY CONNECTIONS

Interactive Whiteboard with **Polygon Sort**
http://alturl.com/n9sg7

Play to Learn with **Shapes Games**
http://pbskids.org/games/shapes/

HISTORICAL CONNECTIONS

The English word "cube" comes from Arabic "k'ab", meaning cube. Chinese, Indian and Islamic scholars were hard at work making contributions to mathematics during Europe's Dark Ages (when intellectual endeavor stagnated). In the 1100's and through the early 1200's, trade expanded the practical need for math. Fibonacci -the first significant mathematician in Europe in more than 1000 years- introduced Hindu-Arabic numerals and geometric vocabulary to Europe.

MATH BACKGROUND - Cubes

A **CUBE** can be folded using 11 different possible nets (patterns for folding). In geometry, a cube is a three-dimensional solid object bounded by six square faces, facets or sides, with three meeting at each vertex. The cube is the only regular hexahedron and is one of the five Platonic solids. The cube is also a square parallelepiped, an equilateral cuboid and a right rhombohedron. It is a regular square prism in three orientations, and a trigonal trapezohedron in four orientations.

Faces	6 Sides
Vertices	8 Corners
Edges	12

MINDBUGS TO NOTE

Many students use cubes to learn multiples of ten. A single cube represents "1." 10 singles makes a row of 10. A row of ten cubes laid side by side becomes "100." "1000" is represented by stacking rows of ten to form a cube.

How do you square a cube? How do you cube a cube? How do you multiply a cube by 10? While using the block system to teach counting, multiples and base ten seems reasonable, beware that children can be confused by the 3-dimensional representation.

Swat the MindBug: When teaching with the cube system, also count using many other options. Bag rice or beans. Use a number line.

RHOMBI'S HOUSE
CURRICULUM CONNECTIONS

COMMON CORE CONNECTIONS

ELA/Literacy
RI.K.1 With prompting and support, ask and answer questions about key details in a text. **W.K.1** Use a combination of drawing, dictating, and writing to compose opinion pieces in which they tell a reader the topic or the name of the book they are writing about and state an opinion or preference about the topic or book. **W.K.2** Use a combination of drawing, dictating, and writing to compose informative/explanatory texts in which they name what they are writing about and supply some information about the topic.

Mathematics
MP.2 Reason abstractly and quantitatively. **MP.4** Model with mathematics. **K.CC.A** Know number names and the count sequence. **K.MD.A.1** Describe measurable attributes of objects, such as length or weight. Describe several measurable attributes of a single object. **1MD.A.1** Describe measurable attributes of objects, such as length or weight. Describe several measurable attributes of a single object.

NEXT GEN SCI CONNECTIONS

ETS1.A: Defining Engineering Problems
A situation that people want to change or create can be approached as a problem to be solved through engineering. Such problems may have many acceptable solutions.

1-ESS1-1, 1-ESS1-2 Patterns in the natural world can be observed, used to describe phenomena, and used as evidence.

ESS2.D: Weather and Climate
Weather is the combination of sunlight, wind, snow or rain, and temperature in a particular region at a particular time. People measure these conditions to describe and record the weather and to notice patterns over time. (K-ESS2-1)

K-ESS2-1. Use and share observations of local weather conditions to describe patterns over time.

UNIT 1: S.T.E.A.M. ACTIVITIES FOR RHOMBI'S HOUSE

Rhombi Audio Download / Video (available December 2014)

UNIT1: Rhombi's House: Science:	Explore wind with anemometers.
UNIT1: Rhombi's House: Technology:	Develop wind-proof barriers.
UNIT1: Rhombi's House: Engineering:	Investigate structures
UNIT1: Rhombi's House: Art & Music:	Create a new picture using shapes.
UNIT1: Rhombi's House: Mathematics:	Draw cubes from various perspective.
UNIT1: Rhombi's House: ELA:	Read and illustrate "Rhombi's House."

RHOMBI'S HOUSE
BUILD S.T.E.A.M. WITH GREAT BOOKS

SCIENCE

Feel the Wind
Arthur Dorros

W is for the Wind
A Weather Alphabet
Dr. Suess

TECHNOLOGY

The Industrial Revolution Build It Yourself
Carla Mooney

Mistakes that Worked
Charlotte Jones Degen

ENGINEERING

How a House Is Built
Gail Gibbons

Olivia Builds a House
Maggie Testa

ART & MUSIC

Museum Shapes
Metropolitan Museum of Art

I Spy Shapes in Art
Lucy Micklethwaite

MATH

Cubes, Cones, Cylinders and Spheres
Tana Hoban

Shapes Around You: 3-D Shapes
Julia Wall

RHOMBI'S HOUSE
STEMVESTIGATION: WEATHER

READ Wind on the Windows *page 60*

DISCUSS What do we already know about wind? Describe wind's motion during various types of weather.

EXPLAIN Wind is air in motion powered by the sun. Sunlight warms the ground in many places. Heat from the ground warms the nearby air. As air grows warmer, it starts to rise. Warm air rises, and cooler air from cooler spots rushes in to fill the space. Wind is described based on where the wind started. "Calm air" moves less than 1 mile per hour. "Hurricanes" cause violent destruction at 74 or more miles per hour.

MOVE Simulate wind. Students wearing warm colors act as the warm air. Students in cool colors act as the cooler air. As the warm group moves out of an area, the cool group moves in waving fans.

INVESTIGATE

How hard is the wind blowing? Build an anemometer to measure wind speed and determine the wind's direction. The anemometer uses four cups that catch the wind and turn the anemometer to measure wind speed. Collect data about wind in at least 7 locations around your school for a month. Plot the wind speed vs. wind direction. Plot wind speed vs. day's temperature. Compare and contrast data for all locations. Use the data analysis to choose a great day for flying kites, picnicking outside, or taking part in a similar class goal.

Obtain five plastic cups. Poke a straw-sized hole through the sides of four cups. Poke a pencil through the bottom of the fifth cup in the center. Ensure all four cups are facing the same direction. Push a straw through two cups at a time, making two pairs. Those straws will also cross through the fifth cup in the center. Place the bottom of the pencil in clay or something sturdy to hold it up. Mark one of the cups to make it easily identifiable. You will count this cup to determine wind speed. Ten turns per minute equals 1 mile-per-hour wind speed. (60 seconds per minute: 60 miles per hour)

ELA Read and Explore "W is for the Wind" A Weather Alphabet.

RHOMBI'S HOUSE
STEMVESTIGATION: DESIGN

TECHNOLOGY

READ   *Accidental Inventions* page 62

DISCUSS   Have you ever made a mistake that ended up with positive results? Like the commercials that show chocolate accidentally dipped in peanut butter, some mistakes end in great inventions.

EXPLAIN   Companies like 3M "solve problems by applying creativity and ingenuity to make life easier." Adhesives like tape and glue are one area in which technology has changed our lives. What adhesives can we name?

MOVE   What items in the classroom / school / playground are held together at seams, sides, joints and corners? What technology is used to keep the items together? (cement, nails, tape, glue, etc.)

INVESTIGATE

Challenge: Investigate ways to connect two pieces of paper such that no wind goes through the seam. Work in teams like professionals so that your creativity is boosted by shared ideas. Each team chooses a different type of paper to investigate. All teams should test at least 4 ways to connect their paper. Get creative. Innovate! What crazy ways could you come up with to keep the wind on one side of the "wall?"

Media: paper, cardboard, cardstock, colored paper, construction paper, foam board, vellum, ??? What else can you find to test?
Adhesive suggestions: masking tape, painter's tape, clear tape, duct tape, glue, paste, paper across the seam, ??? What else can you test?

ELA   On return to the classroom, use the audio from your invention excursion to relive and illustrate the trip.

RHOMBI'S HOUSE
STEMVESTIGATION: BUILDING STRUCTURES

ENGINEERING

READ   *The Tiny House* page 68

DISCUSS   How many students have watched someone sew using a pattern? Has anyone ever folded a cube? What do students already know about the structures in which they live? If you unfolded your house, what do you think it would look like?

EXPLAIN   The simple, square shape of a box house (like most that we live in now) means that all the parts could be unfolded to show a "net" of the building. Nets are used in mathematics to show a shape's faces. Nets are used in science to map the universe!

MOVE   Stretch using the terminology of architecture and a "Rhombi Says" format. "Rhombi says to stretch toward the sky like a skyscraper. Rhombi says to reach deep toward the basement."

INVESTIGATE

Nets are patterns that you can cut and fold to make a model of a solid shape. There are several different ways that you could fold a cube or a pyramid. Unfold several cubes (made from different nets) to trace their patterns. Do the same for pyramids.

Challenge students to draw the nets for their own cubes. Challenge them to trace, cut and fold a pattern to build their own cubes.

GET READY   Use various nets to fold cubes. Use removable tape or painter's tape that allows kids to reopen and reclose the nets you've provided.

RHOMBI'S HOUSE
STEMVESTIGATION: CUBISM

The ARTS

READ — *Cube World* page 66

DISCUSS — Look at a favorite classroom storybook. Do the images look exactly like real objects? If not, what shapes are you able to find in the pictures?

EXPLAIN — Cubism comes from the Latin word cubus, meaning a square or cube. Artists break down objects and people to basic geometric forms. Artists also show objects from more than one angle (front, back, top) in the same painting. Picasso and Braque started the cubist movement in 1908.

MOVE — Choose a cubist painting to reenact as a class, or give each team in the classroom a different picture. Record the reenactments, and challenge other teams to choose the corresponding picture!

INVESTIGATE

Choose magazine ads to cover with geometric shapes. The goal is to end up with a recognizable image using geometric shapes in lieu of the photographic image. Determine the general shape of objects and figures in the ad. Cut 2D geometric shapes to cover objects and figures, adding hand-drawn eyes and lips and ears (chance for a sensory discussion). Add any outstanding features of objects (stripes, dots, etc.).

ELA — Describe the process of creating the "cubist" artwork.

RHOMBI'S HOUSE
STEMVESTIGATION: PERSPECTIVE

READ — Do You Know the Quadrilaterals *page 68*

DISCUSS — What 2D shapes do you see in cubes and cylinders? If you look at a stack of cubes, can you see all the cubes that make up the stack?

EXPLAIN — Google describes *perspective* "as the art of drawing solid objects on a two-dimensional surface so as to give the right impression of their height, width, depth, and position in relation to each other when viewed from a particular point." Perspective changes as the object is turned or moved closer / farther away.

MOVE — Look at a single object from 4 or more different locations. How does the *perspective* change? Does the shape look exactly the same from every location?

INVESTIGATE

Stack blocks for each team with 9 blocks at the base. Stack another 9 blocks at the center, and finish the stack with 9 more blocks. Even though each face of your cube full of cubes is exactly the same, the perspective makes top and side look slanted. Draw the shape. Describe the shape. Create more stacked cubes made of cubes. How many cubes are actually present in the shape? How many cubes are you able to count without removing any layers?

ELA — Extend the discussion without blocks on the table.

RHOMBI'S HOUSE
PLAY WITH SHAPES

SOCK IT TO ME

MATERIALS
Clean disposable sock for each 3D shape
cube, pyramid, cone, cylinder, rectangular prism, sphere
Black marker
stopwatch

SETUP
Place one shape in each clean sock. Number each sock.

INSTRUCT THE PLAYERS
Students teams time each other naming items in the socks. Player 1 feels the shape through the sock and names each shape in turn along with the number listed on the sock. Player 2 uses the stopwatch to record time in seconds. Player 3 double checks the numbers against the shape name on paper to make sure the student names the correct shape. Players will attempt to decrease the number of seconds required to name all objects.

(Discuss why the number of seconds should go down instead of up...)

INSTRUCT THE PLAYERS
Students could easily do this activity in a station
alone or with one teammate.

RHOMBI'S HOUSE

RHOMBI'S ADVENTURES IN 3D
UNIT 1: CUBES

RHOMBI'S HOUSE

Once upon a time, there was Rhombi. Rhombi loved shapes and found them everywhere. She especially loved the squares that made up each face of a cube.

RHOMBI'S HOUSE

Rhombi had just moved to a new town and needed to build a home. Rhombi chose a shape for her house that looked like blocks and sugar cubes.

RHOMBI'S HOUSE

Rhombi knew that she should build a sturdy house to withstand the fall winds, but the last warm days of summer called. The forecast said only partly cloudy, so Rhombi was not worried about the weather.

RHOMBI'S HOUSE

Rhombi hurried to finish her home so she could run and play in the sun. She drew a sketch of her house. Rhombi decided that she could fix it before the weekend. She didn't want to miss the apple-picking party.

RHOMBI'S HOUSE

That night, the weather changed without warning. The air grew colder and blew wildly around the corners of Rhombi's home. Rhombi and Pi walked through the park to watch the wind storm. Pi took pictures but dashed home when the wind gusted.

RHOMBI'S HOUSE

Rhombi ran home. She jumped in surprise as one sharp gust of wind pushed at her house. Another gust of wind pulled at the house. The walls shuddered. The house shivered. So did Rhombi.

RHOMBI'S HOUSE

She stared at all the crispy leaves as they tumbled from the trees. Then, Rhombi stared in amazement as the walls of her house also tumbled to the ground.

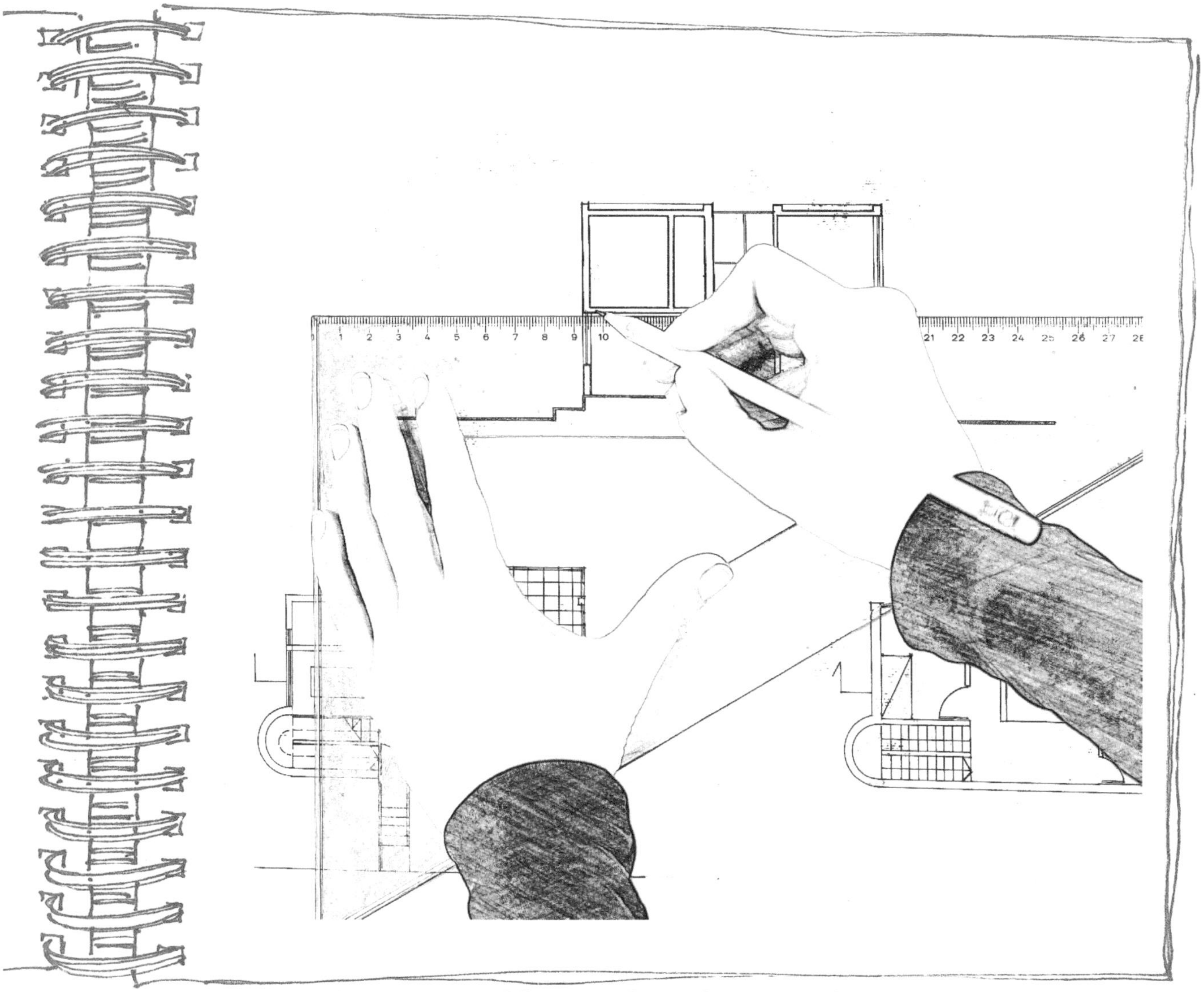

RHOMBI'S HOUSE

Looking at the walls piled on the cold ground, Rhombi decided she needed some help. She looked out beyond the pages of her world and asked, "Designers, can you help me create a new and stronger house?"

RHOMBI'S HOUSE
TALK ABOUT S.T.E.A.M.

How many 2 dimensional square faces make a 3 dimensional cube? **6**

How is a 2D object different from a 3D object? **2 dimensions means the shape has length and width. 3 dimensions means the shape has length, width and height.**

What other objects are shaped like a cube with 6 square faces?
Possible answers include dice, houses, individual keys on a keyboard, old television sets, alarm clocks, storage cubes, moving boxes, and packing boxes.

In what month does summer change to fall?
The Autumnal Equinox in September is considered to be the first day of the Fall Season, when the length of night and day are approximately equal in length.

What other story characters play instead of doing their work?
The cricket in "The Cricket and the Ant" avoids work in summer and is saved by the hard working ants. Two brother pigs in "Three Little Pigs" hurry through building homes. "Little Boy Blue" also leaves his "sheep in the meadows and cows in the corn."

In what season do leaves fall from the trees? **Fall / Autumn**

What does "the house shivered" mean? What makes you shiver?
Its walls moved with the gusty wind. Cold or fear make us shiver.

Does the house shiver for the same reason that Rhombi shivers?
No. The "house shivered" is a metaphor. Rhombi shivers from cold or fear.

What else "tumbled" to the ground? **Leaves tumbled to the ground.**

How would tumbling leaves differ from tumbling walls? (5 senses)
Leaves float and turn in the air, tossed by the wind. Leaves make no noise as they fall. Walls drop with a loud noise, and they shake the ground when they tumble.

What materials will you need in order to build Rhombi's house?
6 squares, tape

RHOMBI'S ADVENTURES IN 3D
UNIT 1 CHALLENGE: RHOMBI'S HOUSE

"Designers, will you create a new, stronger home?"

1. The scale model house walls must be sturdy enough to stand the force of wind simulated by a household hair dryer at highest setting (29mph).
2. The home must measure at least 3inches on each edge.
3. The home must have "L" shaped corners (vertices) of 90°.

Choose squares for your house. Measure to make sure that all 4 sides are the same length. Make sure the corner angles are shaped like an L at 90 degrees.

Use your square tool to trace and cut your own squares for practice. Circle the wall materials you will test.

- construction paper
- single piece of notebook paper
- cardboard
- several pieces of paper glued together
- cardstock paper (scrapbook paper)
- plastic

How will you reinforce the edges? With what materials will you attach each square?

glue
clear tape
masking tape
painter tape
duct tape
electrical tape
folded paper and glue

On the back of this paper, write out the steps you will take to complete your task.

RHOMBOSTEAM 3D SHAPES 2nd Grade   ALL RIGHTS RESERVED ©2014 1O80 EDUCATION, INC.

RHOMBI'S ADVENTURES IN 3D
UNIT 1 RUBRIC: RHOMBI'S HOUSE

TASK — Students will use their knowledge of habitat, house, cube and strength to build a scale model of a playhouse that will not fall apart from the pressure of a hair dryer on high (29mph or higher).

	Content	**Organization**	**Design & Build**
1	• Is well thought out and supports the solution to the challenge or question • Reflects application of critical thinking • Has clear goal that is related to the topic • Is accurate	• Information is clearly focused in an organized and thoughtful manner • Information is constructed in a logical pattern to support the solution	• Make or draw a cube. • Uses squares to create new cubes • Neatly joins corners and edges • Remains cohesive even if blown by the hair dryer • Student adjusts concept in response to dryer test
2	• Supports the solution • Has application of critical thinking that is apparent • Has no clear goal • Has some factual errors or inconsistencies (i.e. wheels do not rotate around axle)	• Project has a focus but might stray from it at times (more concerned with form than function) • Information appears to have a pattern, but the pattern is not consistently carried out in the project	• Identifies a square but does not combine them to make a cube • Edge seams are mis-joined and uneven • Student is unable to adjust concept following dryer test
3	• Provides inconsistent information for solution • Has no apparent application of critical thinking • Has no clear goal • Has significant factual errors, misconceptions, or misinterpretations	• Content is unfocused and haphazard • Information does not support the solution to the challenge or question • Information has no apparent pattern	• Does not draw/make 2D shapes • Does not know to join squares to make a cube • Edge seams are missing or ragged • Student project falls apart and student is unable to make changes

CUBE NET
PRINT AND LAMINATE FOR USE DURING STEMVESTIGATIONS.

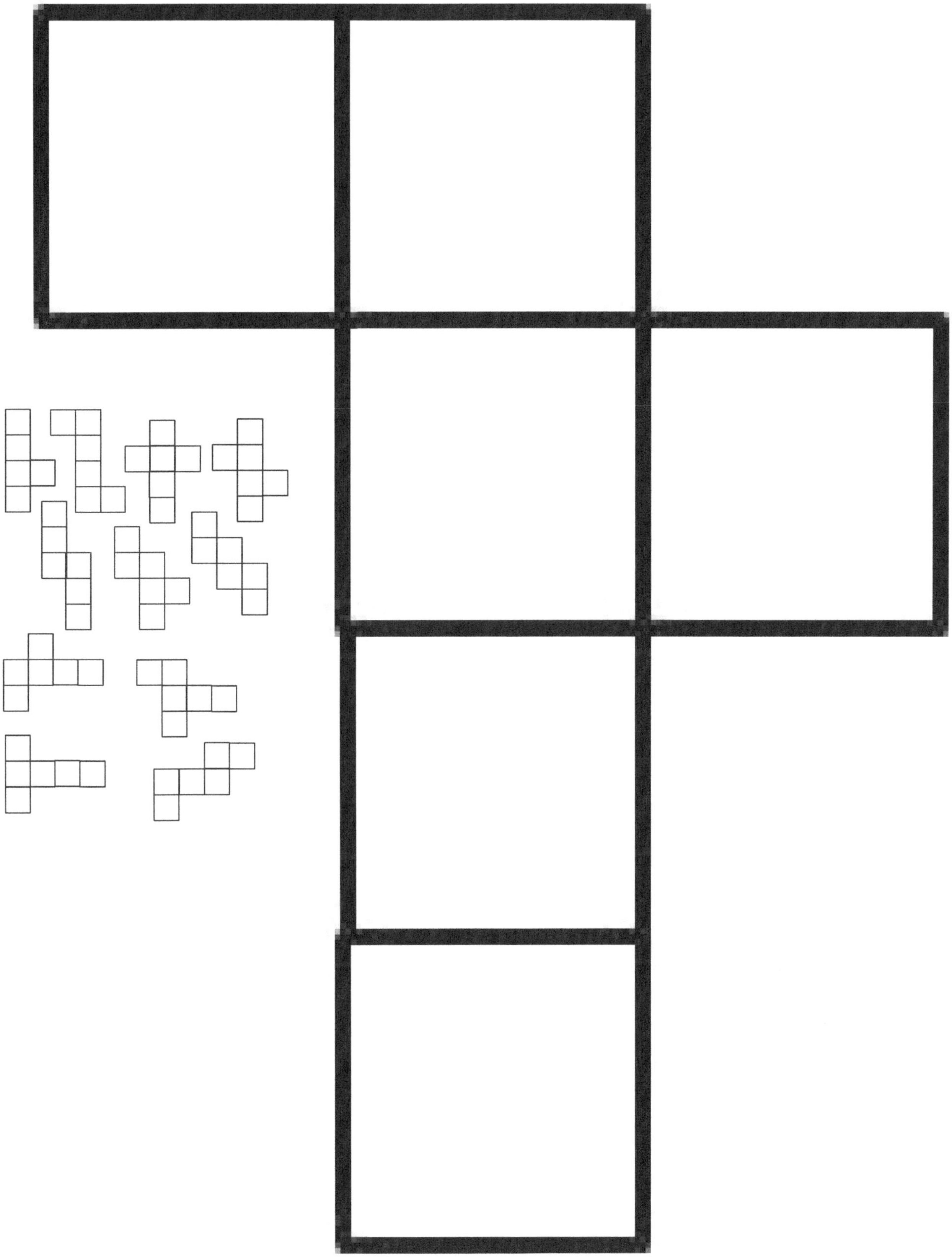

RHOMBOSTEAM 3D SHAPES 2nd Grade   ALL RIGHTS RESERVED ©2014 1O80 EDUCATION, INC.

UP ON THE ROOF

RHOMBI'S ADVENTURES IN 3D
UNIT 2: PYRAMIDS

RHOMBI'S ADVENTURES IN 2D
UNIT 2: UP ON THE ROOF

UP ON THE ROOF

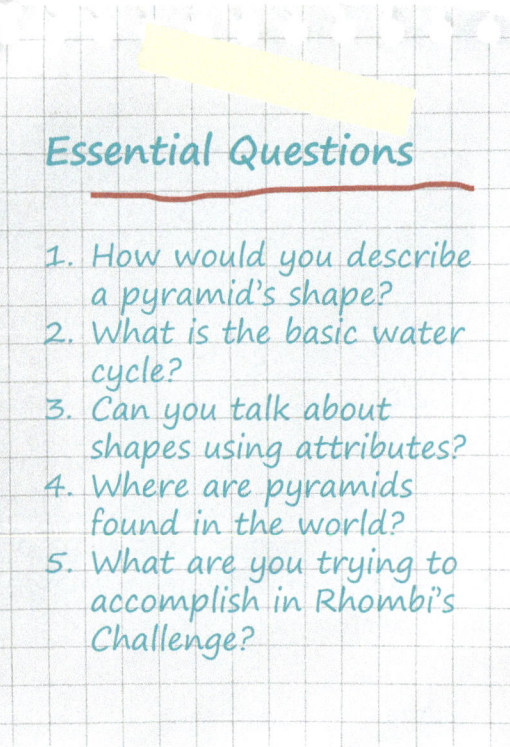

Essential Questions

1. How would you describe a pyramid's shape?
2. What is the basic water cycle?
3. Can you talk about shapes using attributes?
4. Where are pyramids found in the world?
5. What are you trying to accomplish in Rhombi's Challenge?

Focus on triangles, squares, and pyramids while reinforcing cubes. Learn water cycles. Explore materials with sun baked images. Decide the most efficient way to build a pyramid net. Play with

Students will apply basic measurement, understanding of number, place value, data collection, and data application in problem solving. Student will use two-dimensional shapes to create a cube and a pyramid.

Rhombi's playhouse looks great, but the roof is leaking. Students will apply new understanding of shapes, weather, materials, water cycles, and pyramids to design and construct a waterproof roof.

ENDURING UNDERSTANDING

Two dimensional shapes create new shapes.

"Halves" means two equal parts."

Circles and rectangles break into equal parts.

Breaking circles or rectangles into more equal parts means that the parts will be smaller.

Three objects may be put in order from longest to shortest by comparing their lengths.

The length of an object may be discussed using whole numbers.

A smaller object may be used as a measurement tool when measuring longer lengths.

Larger numbers consist of tens and ones.

Organizing information helps in discussion.

freezing
pass from the liquid to the solid state by loss of heat

thawing
change from a frozen solid to a liquid by gradual warming

pyramid
solid figure with a polygonal base and triangular faces that meet at a common point

waterproof
to make unaffected by water

UP ON THE ROOF
CURRICULUM CONNECTIONS

Rhombi needs a pyramid playhouse roof to withstand rainy weather.

ART CONNECTIONS - Ancient Egypt

PYRAMIDS

While we know that the stone for the pyramids was quarried, transported and cut from the nearby Nile, we still cannot be sure just how the massive stones were then put into place. While stone was generally reserved for tombs and temples, sun-baked mud bricks were used in the construction of Egyptian houses, palaces, fortresses, and town walls.

TECHNOLOGY CONNECTIONS

Play to Learn
Interactive Polygons
http://www.learner.org/interactives/geometry/3d_pyramids.html

HISTORICAL CONNECTIONS

PYRAMIDS are present in architecture dating to earliest recorded history. Examples include:
• Egypt- Great Pyramids of Giza
• the Americas - Pyramid of the Sun / of the Moon

According to History.com, the Americas contained "more pyramid structures than the rest of the planet combined." Mayans, Aztecs and Olmecs built pyramids for deities and burial. Latin American pyramids include the Pyramid of the Sun and the Pyramid of the Moon at Teotihuacán in central Mexico.

MATH BACKGROUND - Pyramid

In geometry, a SQUARE PYRAMID is a pyramid having a square base. If the apex is perpendicularly above the center of the square, it will have what is called C4v symmetry.

A Johnson solid is one of 92 strictly convex regular-faced polyhedra, but which is not uniform, i.e., not a Platonic solid, Archimedean solid, prism or antiprism. They are named by Norman Johnson who first enumerated the set in 1966.

Requires	4 triangles & 1 Square
Edges	8
Vertices	5

MINDBUGS TO NOTE

A single sheet of paper will not register on a digital scale or change the balance on a balance scale. Students mistakenly believe that paper weighs nothing.

MindBugs in Numeracy

Prove that paper has weight.

Discover the weight of a single piece of paper by weighing a stack of 10-20 pieces. On a number line, count back by 10s or 20s to determine the weight of a single piece of the stack.

Swat the MindBug: Weigh multiple items, and use a number line to "divide."

UP ON THE ROOF
CURRICULUM CONNECTIONS

COMMON CORE CONNECTIONS

ELA/Literacy
RI.K.1 With prompting and support, ask and answer questions about key details in a text. **W.K.1** Use a combination of drawing, dictating, and writing to compose opinion pieces in which they tell a reader the topic or the name of the book they are writing about and state an opinion or preference about the topic or book. **W.K.2** Use a combination of drawing, dictating, and writing to compose informative/explanatory texts in which they name what they are writing about and supply some information about the topic.

Mathematics
MP.2 Reason abstractly and quantitatively. **MP.4** Model with mathematics. **K.CC.A** Know number names and the count sequence. **K.MD.A.1** Describe measurable attributes of objects, such as length or weight. Describe several measurable attributes of a single object. **MD.A.1** Describe measurable attributes of objects, such as length or weight. Describe several measurable attributes of a single object.

NEXT GEN SCI CONNECTIONS

ETS1.A: Defining Engineering Problems
A situation that people want to change or create can be approached as a problem to be solved through engineering. Such problems may have many acceptable solutions.

1-ESS1-1,1-ESS1-2 Patterns in the natural world can be observed, used to describe phenomena, and used as evidence.

ESS2.D: Weather and Climate
Weather is the combination of sunlight, wind, snow or rain, and temperature in a particular region at a particular time. People measure these conditions to describe and record the weather and to notice patterns over time. (K-ESS2-1)

K-ESS2-1. Use and share observations of local weather conditions to describe patterns over time.

UNIT 2: S.T.E.A.M. ACTIVITIES FOR "UP ON THE ROOF"

Rhombi Audio Download / Video (available August 2014)

UNIT 2:	Rhombi's Up on the Roof: Science:	Explore evaporation and condensation.
UNIT 2:	Rhombi's Up on the Roof: Technology:	Visit Egyptian pyramids in a virtual field trip.
UNIT 2:	Rhombi's Up on the Roof: Engineering:	Investigate gravity and structures with pyramids.
UNIT 2:	Rhombi's Up on the Roof: Art & Music:	Develop a clay brick recipe.
UNIT 2:	Rhombi's Up on the Roof: Mathematics:	Use clay bricks to build sturdy structures.
UNIT 2:	Rhombi's Up on the Roof: ELA:	Read and illustrate "Up on the Roof."

UP ON THE ROOF
BUILD S.T.E.A.M. WITH GREAT BOOKS

SCIENCE

Suns
Franklyn M. Branley

Water Cycle Videos
http://www.youtube.com/user/makemegenius/videos

TECHNOLOGY

Water Cycle Video
http://alturl.com/2vju3

Explore a Pyramid Interactive Experience
http://education.nationalgeographic.com/education/kd/?ar_a=5

More information and Interactive Adventures
http://www.nationalgeographic.com/pyramids/khufu.html

ENGINEERING

How
Gail Gibbons

Look
Scot Ritchie

ART & MUSIC

If You Lived Here: Houses of the World
by Giles Laroche

Matisse for Kids
http://www.artbma.org/flash/f_conekids.swf

MATH

Cubes
Tana Hoban

Teacher Interactive for White Board - Symmetry Flip / Rotate
http://www.teacherled.com/resources

UP ON THE ROOF
STEMVESTIGATION: WATER CYCLE

READ   **Water Is as Water Does** *page 72*

DISCUSS   Where can you find water? What happens to rain puddles once clouds drift away to reveal the sun?

EXPLAIN   Water on Earth is always changing. Its repeating changes make a cycle. As water goes through its cycle, it can be a solid (ice), a liquid (water), or a gas (water vapor). Ice can change to become water or water vapor. Water can change to become ice or water vapor. Water vapor can change to become ice or water.

MOVE   Be the water! Play the rain game. Sit cross-legged on the ground in a circle. The class pats hands on legs or drums the ground. Simulate a rain storm. Wind is pushed ahead of the storm. Rain falls softly, increasing in strength (slap the ground harder and harder). Thunder booms (Teacher claps hands together.) Rain decreases in strength until the sound dies away altogether.

INVESTIGATE

Choose 4-6 different surfaces to test: include something porous like paper and something nonporous like plastic. Give each team of students responsibility for one surface. Place 5 drops of water on each surface. Measure the width of the puddle. Place the test surfaces in a sunny window or under a lamp. Make a coordinate graph showing the hour of the day vs. width of each puddle.

ELA   Each hour of the day, take digital photos of the surfaces. Make a slide handout in PowerPoint.

UP ON THE ROOF
STEMVESTIGATION: PYRAMIDS

TECHNOLOGY

READ   The Case of the Disappearing Sand *page 74*

DISCUSS   What do students already know about the pyramids? How wide and tall were the Egyptian pyramids? How long did they take to build? What was inside them?

EXPLAIN   Pyramids of Khufu at Giza were built thousands of years ago in Egypt. The Pyramid at Giza stands 481 feet tall (147 meters). This pyramid is a square pyramid because its base is a square and all 4 triangular faces are equilateral triangles.

MOVE   Can your class (or team) make a human pyramid? Walk the length of the pyramid's height (481 feet).

INVESTIGATE

Visit a real pyramid in the virtual world. Take a web-based field trip through the Pyramid at Giza.

Explore a Pyramid Interactive Experience
http://education.nationalgeographic.com/education/kd/?ar_a=5

Build a Pyramid Online
http://www.bbc.co.uk/history/ancient/egyptians/launch_gms_pyramid_builder.shtml

ELA   Create a comic strip to describe how the pyramid was built in your interactive experience.

UP ON THE ROOF
STEMVESTIGATION: ROOFTOPS AND HEIGHT

ENGINEERING

READ Circle, Triangle, Square *page 76*

DISCUSS Which of the buildings have pointed rooftops? How tall are the "pyramids" or "cones?" What other names can you think of for the tall upper part of a building? Why might buildings need pointed tops? (wind, rain)

EXPLAIN Some of the tops of buildings are cones with circle bases and pointed top. Some buildings are topped with a pyramid shape (square base and triangular faces).

MOVE Students lie down head to feet to make a human ruler. How many students would need to stand on each other's shoulders to touch the tallest part of the Chrysler Building? According to **CNN: 25 Great Skyscrapers: Icons of Construction**, the spire rises 186 feet feet above the main building.

INVESTIGATE

How tall can a top (apex) rise before the building topples the pyramid? What steps can be taken to keep the building stable no matter how tall the spire rises?

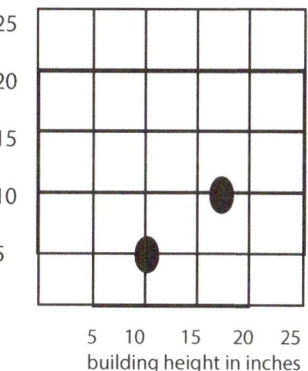

Challenge: Build the tallest pyramid spire possible before the building topples. Measure the height of the building for each attempt. Measure the height of the spire for each attempt. Graph building height vs. spire height.

ELA Write about the "story" your graph tells.

UP ON THE ROOF
STEMVESTIGATION: MIXTURES & EVAPORATION

READ — I Love Homes *page 78*

DISCUSS — Have you ever made mud pies? Clay figures? How long does your creation take to dry? Do all clays dry?

EXPLAIN — Around the world, bricks are made of local materials. From clay to soil and dirt, people use the tools at hand to build homes. Clay or dirt alone will crumble. Adding grass, straw or string keeps the brick together. Sculptors also use clay mixtures to create works of art. Women of the Southwestern Native American tribes normally created pots, and the men decorated them with designs like simple geometric patterns.

MOVE — Look for clay, straw and soils around the building that might be used to make tiny pyramid.

INVESTIGATE

Make the Mixture: Bricks may be made when clay soil is mixed with straw. Mix soil with water until it becomes quite thick. Add the straw. How much straw should be used? How much water? What is the best mixture to make a clay shape that dries without crumbling? You might use candy or cookie moulds, or just form figures with fingers. Make a scale model of a pyramid once you've determined the best mixture of clay, dirt, straw... Leave the structure out in full sun for 2 days to bake the clay. Decorate!

ELA — Watch BBC instructors build bricks!
http://www.bbc.co.uk/learningzone/clips/how-are-bricks-made/13460.html

UP ON THE ROOF
STEMVESTIGATION: EROSION

READ   The Camel and the Pyramid *page 80*

DISCUSS   Why do people use pyramids? If rain is the main reason, why would pyramids be used in Egypt? Wind is also a factor: wind has less surface to push when the building narrows to a point. Less wind, less erosion.

EXPLAIN   The Egyptian pyramids are more than 3000 years old. The structures are not as sharp as they once were. Wind and rain break down materials over time.

MOVE   Use a sheet, rod, square base, and heavy-duty binder clips to make a pyramid tent. Teamwork is required to build the structure! Make a hole in the center of your square base (cardboard works well). Insert the tall rod, and have one student hold it steady. Drape a sheet over the rod. Secure sides / corners with binders clipped to the base.

INVESTIGATE

Make a mixed clay pyramid. Once baked in the sun, measure and record the length of sides and height of the pyramid. Students that are too young to measure with a ruler should place a piece of paper or cardboard behind the structure, and trace its original outline. Place the pyramid in a tray. Set up a drip system or "rain" on the structure with a watering can containing a measured volume of water. Record the amount of water that will -eventually- be dripped onto the pyramid. How much water over some length of time will begin to erode the building?

Repeat the experiment with wind (hair dryer on high or leaf blower).

Alternate Activity: Leave the pyramids in the sun. Knowing how much rain has fallen and how many days have passed will yield the same data.

UP ON THE ROOF
PLAY WITH SHAPES

LEARN TO JUGGLE

MATERIALS
soft round balls (3 per student or enough for a single station)

SETUP
Head outside, or clear desks for room to juggle.

INSTRUCT THE PLAYERS
Ancient Egyptian art shows images of Egyptians juggling balls! You can learn to do this, too.

INSTRUCT THE PLAYERS
Students could easily do this activity in a station alone or with one teammate. You'll need to master one level of juggling before moving to the next level.

1. Toss one ball from one hand to the other. Practice until you can toss the ball as high as your eyes.
2. Hold one ball in your weaker hand. Hold a second ball in your stronger hand. Toss the strong hand ball, and -just before you catch it in your weak hand- move the weak hand ball to the strong hand (switch).
3. Repeat above but toss the second ball up and over instead of switching hands.
4. When you are comfortable with 2 balls, add a third.

source   http://learnhowtojuggle.info/the-basics/

UP ON THE ROOF

**RHOMBI'S ADVENTURES IN 3D
UNIT 2: PYRAMIDS**

UP ON THE ROOF

Once upon a time there was Rhombi. She lived in a sturdy new house shaped like a cube. Her home was strong and safe. Autumn winds had not been able to push it down.

UP ON THE ROOF

Rain was a problem, though. Winter rain and piles of snow were making pools of water on the flat rooftop.

UP ON THE ROOF

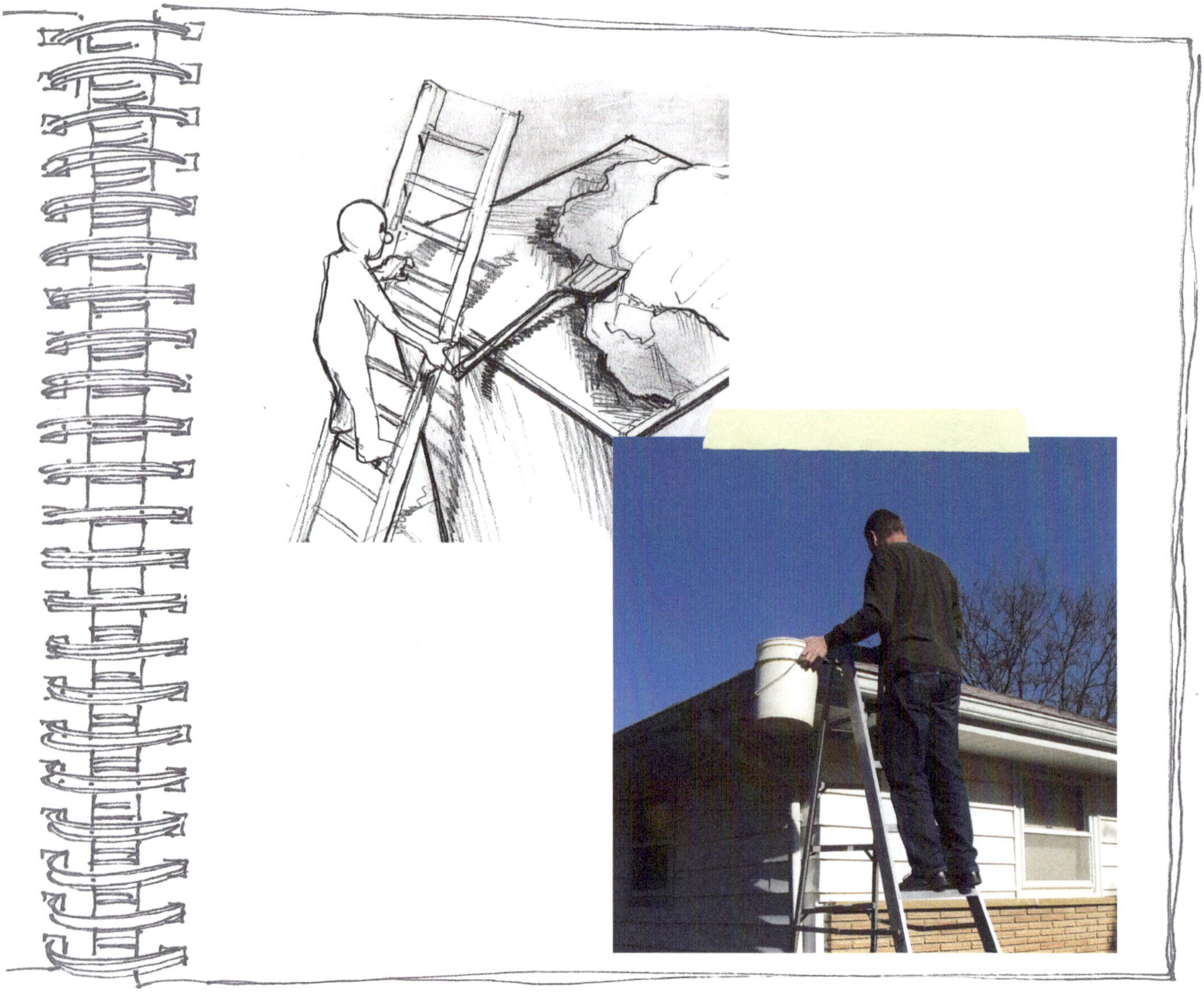

Rhombi's friend climbed a ladder onto the rooftop to investigate. Drip. Drip. Drip. Freezing water and thawing ice had caused a hole in the roof!

UP ON THE ROOF

The water had started to drip, drip, drip through the ceiling and into her home. If she didn't find a way to change the roof, her strong new home would be full of water.

UP ON THE ROOF

Rhombi stood outside with her hot chocolate. She looked at her wonderful house and wondered what to do. She asked Radius and Pi for advice when they visited. The twins thought a pyramid might work.

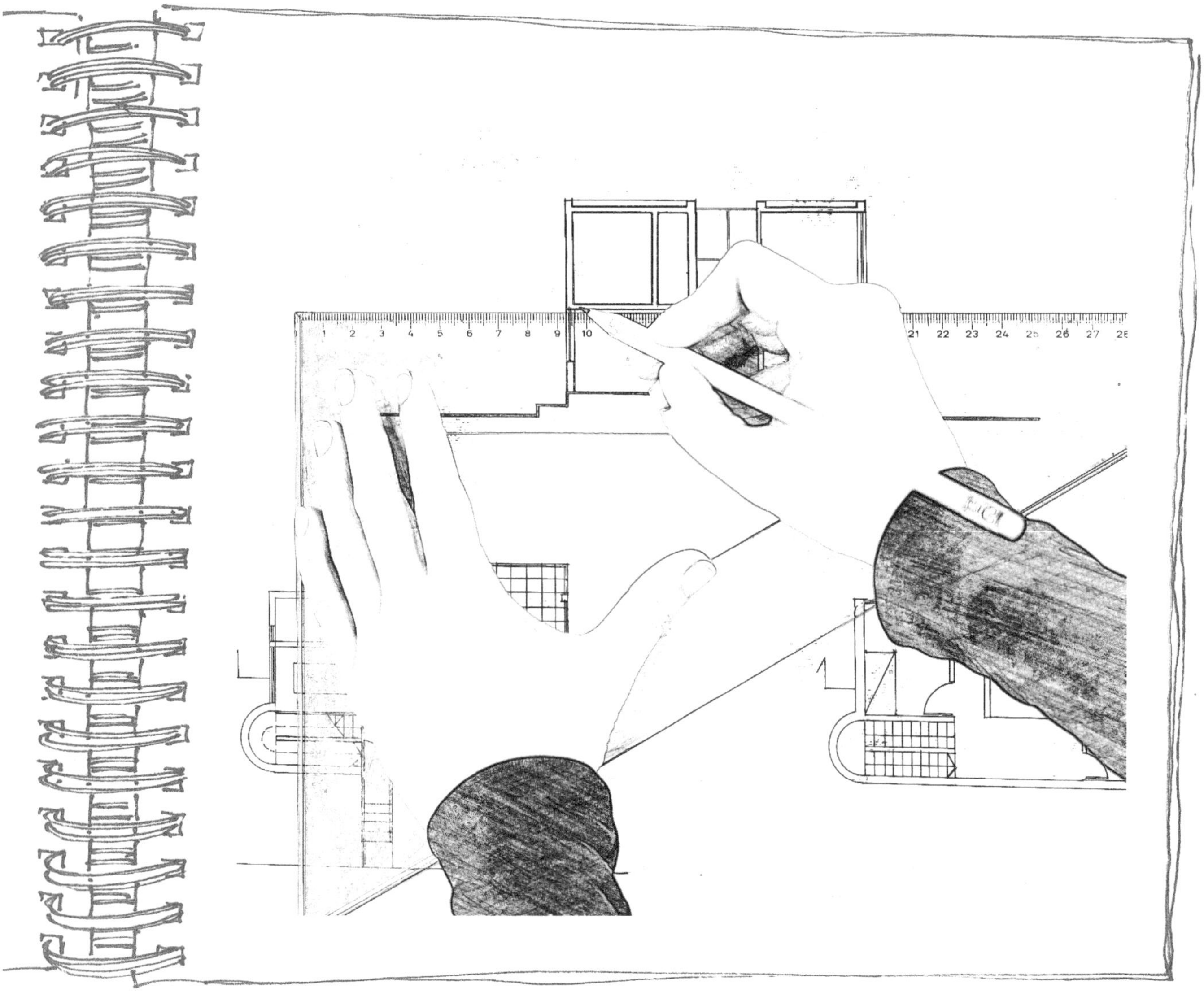

UP ON THE ROOF

Rhombi decided to ask for help. She looked out beyond the pages of her world and asked, "Designers, can you help me create a waterproof roof for my house?"

UP ON THE ROOF
TALK ABOUT S.T.E.A.M.

How many 2 dimensional square and triangular faces make a 3 dimensional cube?
6

How is a 2D object different from a 3D object?
2 dimensions means the shape has length and width. 3 dimensions means the shape has length, width and height.

What is another way to say "autumn?
Fall

What is the average rainfall in your area for the month of February?
Answers will vary.

How are freezing water and thawing ice related?
Frozen water is called ice. Ice melts - or thaws- to make liquid (water).

How would rain get from the roof to the ceiling of the rooms below?
The roof is flat. The underside of the house's roof is also the ceiling of Rhombi's room.

Why is Rhombi wearing rain boots and carrying an umbrella?
She wants to dress appropriately for the rainy weather.

About what is Rhombi wondering?
She wonders what to do to fix her leak.

How could Rhombi change the shape of her flat roof?
She could make it pointy or tent-like.

Why would Rhombi want to change the shape of her roof?
The water is collecting on her flat roof. If she makes a slanted roof, the water will slide off and not cause puddles or leaks through the ceiling.

What materials will you need in order to build Rhombi's house?
1 square base and 4 equilateral triangles

RHOMBI'S ADVENTURES IN 3D
UNIT 2 CHALLENGE: UP ON THE ROOF

"Will you use triangles to design a pyramid-shaped roof for my playhouse?"

Draw Rhombi's cube house without the new roof.	Draw Rhombi's house with a new roof.

Draw each of the pieces you will need to complete the challenge.

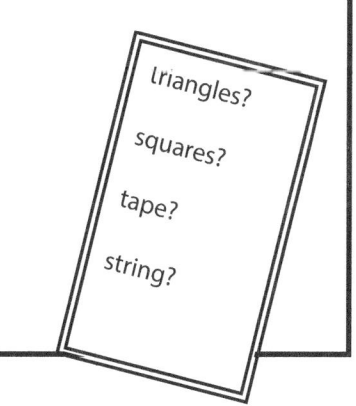

triangles?
squares?
tape?
string?

On the back of this paper, trace the net you will create.

RHOMBI'S ADVENTURES IN 3D
UNIT 2 RUBRIC: UP ON THE ROOF

TASK: Students will use their knowledge of habitat, house, pyramids and strength to build a scale model of a roof that will allow water to fall from the roof without puddling.

	Content	**Organization**	**Design & Build**
1	• Is well thought out and supports the solution to the challenge or question • Reflects application of critical thinking • Has clear goal that is related to the topic • Is accurate	• Information is clearly focused in an organized and thoughtful manner • Information is constructed in a logical pattern to support the solution	• Makes or draw a pyramid • Uses triangle and square to make a pyramid • Corners and edges are neatly joined • Remains cohesive even if "rained" on… • Student adjusts concept in response to "rain" test
2	• Supports the solution • Has application of critical thinking that is apparent • Has no clear goal • Has some factual errors or inconsistencies (i.e. wheels do not rotate around axle)	• Project has a focus but might stray from it at times (more concerned with form than function) • Information appears to have a pattern, but the pattern is not consistently carried out in the project	• Identifies triangle and square but does not combine them to make a pyramid • Edge seams are mis-joined and uneven • Student is unable to adjust concept following "rain" test
3	• Provides inconsistent information for solution • Has no apparent application of critical thinking • Has no clear goal • Has significant factual errors, misconceptions, or misinterpretations	• Content is unfocused and haphazard • Information does not support the solution to the challenge or question • Information has no apparent pattern	• Does not draw/make a pyramid • Does not join triangles and squares to make pyramid • Edge seams are missing or ragged • Student project falls apart and student is unable to make changes

PYRAMID NET
PRINT THIS NET FOR USE DURING STEMVESTIGATIONS.

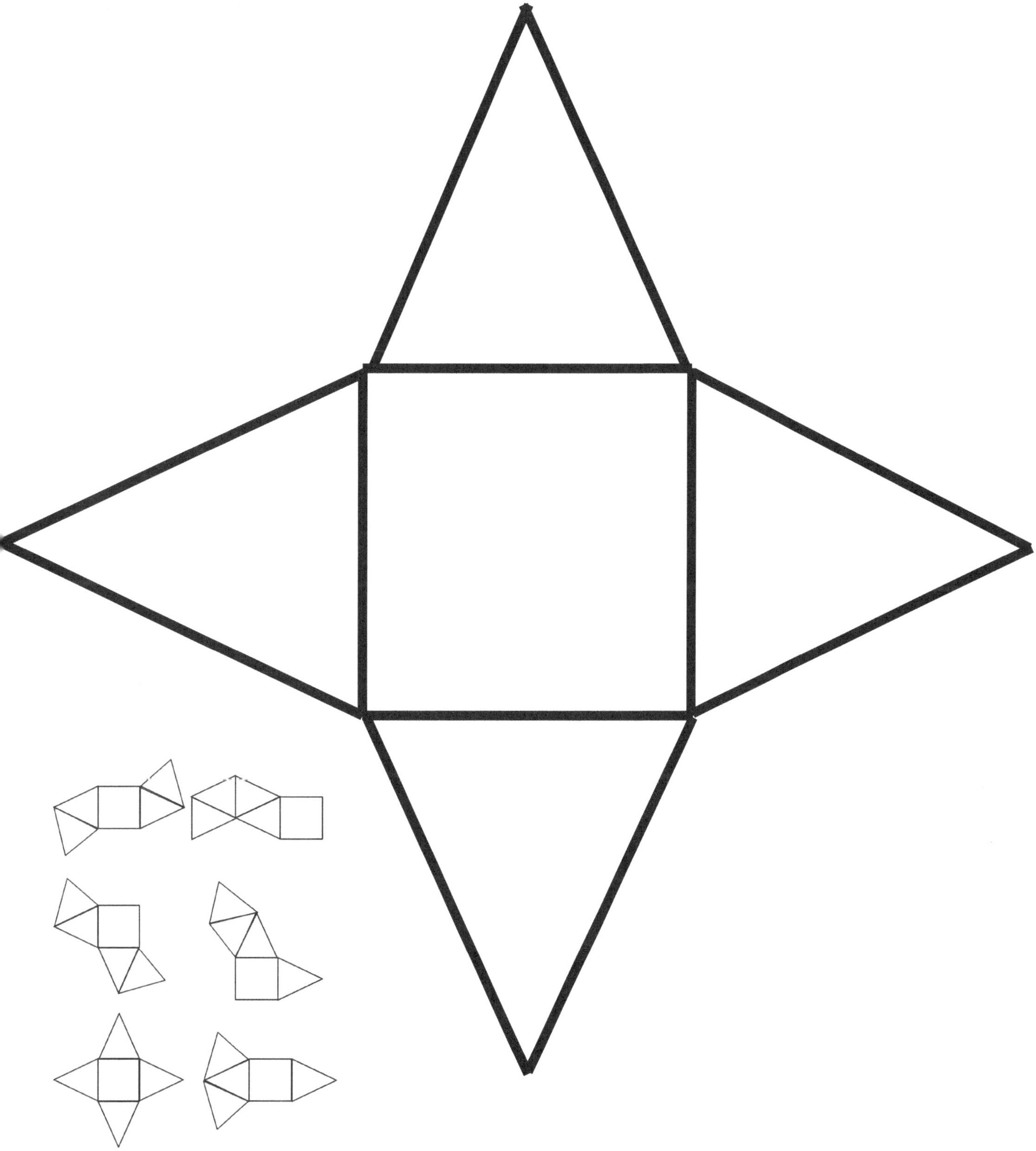

RHOMBOSTEAM 3D SHAPES 2nd Grade   ALL RIGHTS RESERVED ©2014 1O80 EDUCATION, INC.

NEW HOME FOR PET

RHOMBI'S ADVENTURES IN 3D
UNIT 3: HABITATS

RHOMBI'S ADVENTURES IN 3D
UNIT 3: NEW HOME FOR PET

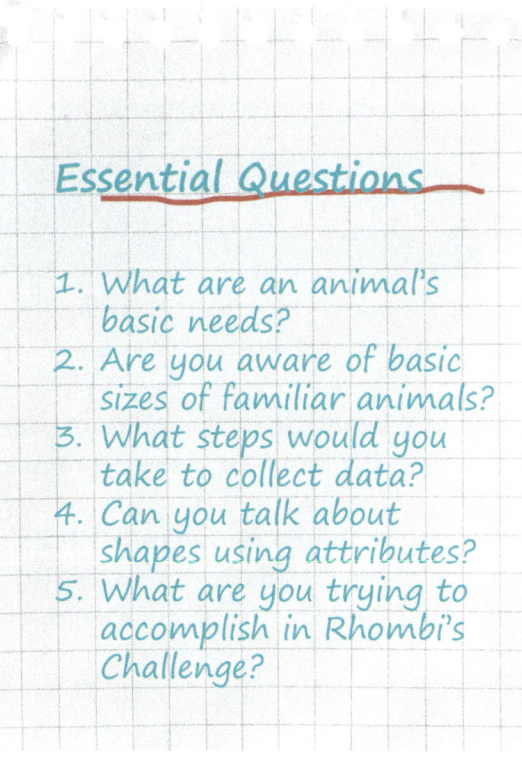

Essential Questions

1. What are an animal's basic needs?
2. Are you aware of basic sizes of familiar animals?
3. What steps would you take to collect data?
4. Can you talk about shapes using attributes?
5. What are you trying to accomplish in Rhombi's Challenge?

ENDURING UNDERSTANDING

Two dimensional shapes create new shapes.

"Halves" means two equal parts."

Circles and rectangles break into equal parts.

Breaking circles or rectangles into more equal parts means that the parts will be smaller.

Three objects may be put in order from longest to shortest by comparing their lengths.

The length of an object may be discussed using whole numbers.

A smaller object may be used as a measurement tool when measuring longer lengths.

Larger numbers consist of tens and ones.

Organizing information helps in discussion.

NEW HOME FOR PET

Focus on combining two-dimensional shapes to create three-dimensional shapes for a purpose. Investigate basic needs. Discover the natural habitat of the South American guinea pig. Develop a

Reinforce for students that may have missed the mark in Units 1 and 2. Accelerate for students ready to take on a challenge. Combine cubes and pyramids.

Rhombi welcomes a new pet into her home. She and Pet trip all over each other, and Rhombi decides that her tiny home needs a special room for Pet. Students design and build the scale model for a pet habitat (pet of their choice).

NEW HOME FOR PET
CURRICULUM CONNECTIONS

Rhombi's Pet needs a place of her own.

ART CONNECTIONS - Animals in Art

REALISM TO CUBISM
From the realistic watercolors of Albrecht Dürer to the cubist oils of Franz Marc, animals have been a popular topic of art through the ages. Work with the school media specialist and art teacher to incorporate drawing, painting and sculpture on the subject of animals in art. Visit this great URL for more information and examples:
www.artcyclopedia.com/subjects/Animals.html

TECHNOLOGY CONNECTIONS

Art Projects for Kids: Animal Habitat Sketches
http://www.artprojectsforkids.org

New York Zoo's Website: Create Your Wild Self
http://buildyourwildself.com/

HISTORICAL CONNECTIONS

ANIMALS AS PETS
While scientists are still unable to pinpoint the exact location, they can say with a fair amount of certainty that the "gray wolf" was the first animal to become man's best friend. When humans began settling for longer periods of time, they also started dumping refuse in piles outside the camps. Wild wolves were drawn in and eventually became familiar with humans. Through many generations, humans bred the animals for guard duty and hunting. Dogs have many jobs now, including that of "best friend."

MATH BACKGROUND - Height

In geometry, height is defined as the measurement from an object's base to its top.

In a pyramid, height would be measured from the center of the base to the pinnacle of the pyramid.

In a cube, the height should be measured from base of one edge to the top of the same edge (to avoid accidentally slanting to one side or another).

MINDBUGS TO NOTE

When placed on a scale, a single sheet of paper will not register on a digital scale or change the balance on a balance scale. Students mistakenly believe that paper weighs nothing.

MindBugs in Numeracy

Prove that paper has weight.

Discover the weight of a single piece of paper by weighing a stack of 10-20 pieces. On a number line, count back by 10s or 20s to determine the weight of a single piece of the stack.

Swat the MindBug: Weigh multiple items, and use a number line to "divide."

NEW HOME FOR PET
CURRICULUM CONNECTIONS

COMMON CORE CONNECTIONS

ELA/Literacy
RI.K.1 With prompting and support, ask and answer questions about key details in a text. **W.K.1** Use a combination of drawing, dictating, and writing to compose opinion pieces in which they tell a reader the topic or the name of the book they are writing about and state an opinion or preference about the topic or book. **W.K.2** Use a combination of drawing, dictating, and writing to compose informative/explanatory texts in which they name what they are writing about and supply some information about the topic.

Mathematics
MP.2 Reason abstractly and quantitatively. **MP.4** Model with mathematics. **K.CC.A** Know number names and the count sequence. **K.MD.A.1** Describe measurable attributes of objects, such as length or weight. Describe several measurable attributes of a single object. **MD.A.1** Describe measurable attributes of objects, such as length or weight. Describe several measurable attributes of a single object.

NEXT GEN SCI CONNECTIONS

ETS1.A: Defining Engineering Problems
A situation that people want to change or create can be approached as a problem to be solved through engineering. Such problems may have many acceptable solutions.

1-ESS1-1,1-ESS1-2 Patterns in the natural world can be observed, used to describe phenomena, and used as evidence.

ESS2.D: Weather and Climate
Weather is the combination of sunlight, wind, snow or rain, and temperature in a particular region at a particular time. People measure these conditions to describe and record the weather and to notice patterns over time. (K-ESS2-1)

K-ESS2-1. Use and share observations of local weather conditions to describe patterns over time.

UNIT 3: S.T.E.A.M. ACTIVITIES FOR NEW HOME FOR PET

Rhombi Audio Download / Video (available December 2014).

UNIT 3: New Home for Pet: Explore habitats with a focus on science.
UNIT 3: New Home for Pet: Explore attributes with a focus on technology.
UNIT 3: New Home for Pet: Explore columns with a focus on engineering.
UNIT 3: New Home for Pet: Explore patterns with a focus on art.
UNIT 3: New Home for Pet: Explore scale and size with a focus on mathematics.

UNIT 3: New Home for Pet: ELA: Read and illustrate "PET FINDS A HOME."

NEW HOME FOR PET
BUILD S.T.E.A.M. WITH GREAT BOOKS

SCIENCE

Fruits and Vegetables from A to Z
By Lois Ehlert

Classroom Library:
favorite animals
classroom pets
habitats
needs of living things

TECHNOLOGY

Joe the Gentle Giant and the Badger's Home
by Wayne Stripling

Classroom Library:
animal classification
adaptation

ENGINEERING

Flower Tower oF Power
by Jeannie Ruiz

Classroom Library:
towers
civil engineers

ART & MUSIC

Meet Me at the Art Museum
by David Golden

Classroom Library:
M.C. Escher
tesselations
patterns in nature

MATH

Life-Size Farm:
Teruyuki Komiya

Classroom Library:
measurement
small animals, medium animals, large animals

NEW HOME FOR PET
STEMVESTIGATION: HABITATS

READ   **Davis Finds a Purrl. Purrl Finds a Home.** *page 84*

DISCUSS   Look at the book covers. What makes each animal unique? What things do the animals have in common?

EXPLAIN   Animals need air, food, water, and a safe place to raise young. Pets also need to breath, eat, drink and feel safe in their homes. As the caretaker, you are responsible for providing all these things needed for living things to survive and stay healthy.

MOVE   March through the classroom or playground chanting, "food, water, air and home." As students return to seats, each student should repeat the list to an instructor.

INVESTIGATE

Students bring stuffed animals or toy pets to class. Students will use construction paper, scissors, tape, art materials and cardboard scraps to design and build habitats for their pets. Present the animal's geometric habitat during a show-and-tell using appropriate vocabulary.

ELA   Add an additional technology element with an interview booth. Hang a sheet on the wall. Students make their presentations to a video camera. As everyone enjoys party food, play the presentations for the group!

SCIENCE

NEW HOME FOR PET
STEMVESTIGATION: ATTRIBUTES

TECHNOLOGY

READ   Rhombi Lived in a Zoo *page 86*

DISCUSS   What do we know about animals and their habits? Do fish like to sunbathe on a hot rock? Do dogs make nests in trees to protect their eggs? An animal's body gives clues to where the animal lives and what it eats.

EXPLAIN   Animal attributes match food sources and homes. Example: Animals that swim need fins and gills. Animals with wings can live in a tree or rooftop. Furred animals are able to survive colder weather. Scaled animals are protected on hot surfaces.

MOVE   Call out a habitat, and have students become animals that can survive and thrive in that environment.

INVESTIGATE

The New York Zoo has a great website on which students may mix-and-match to create their own "wild selves." Have students create a list of what they want to eat, where they want to live and what they want to be able to do. Then visit the website and create a "self" that fits the description they've created.

New York Zoo's Website
http://buildyourwildself.com/

ELA   Design a "bio" card for the "self" created at the zoo's site. How will the card be organized to show the following information; name, height, weight, description, food source, habitat, and favorite activities.

NEW HOME FOR PET
STEMVESTIGATION: FOUNDATIONS

READ It All Adds Up *page 88*

DISCUSS Structures are built from the bottom up, a lot like the 3D piece in the story. Every strong building begins with a solid, sturdy foundation.

EXPLAIN Structures can hold more weight when the load is shared with other shapes. The load is shared when a flat "floor" is secured above the support shapes. The floor spreads out weight, so there isn't too much weight on any one cup.

MOVE Everybody, drop for some push-ups. It's the body forming a bridge!

INVESTIGATE

Place one cardboard square on the floor, and a single cup upside down on the square. Place the second square on top. Add weight until the cup bends or breaks. How much weight did the one cup support? Try two cups and retest. Try three cups and retest. Add your data to a class graph with cups on the x axis and weight on the y axis.

Using the data collected, decide how many cups you think you'll need to hold a student's weight. Once your structure is built, ask a teacher to help you step onto the "floor." How few cups will hold your weight?

ELA Tell the story from the cup's point of view!

NEW HOME FOR PET
STEMVESTIGATION: TESSELATIONS

The ARTS

READ — M.C. Escher's World *page 90*

DISCUSS — What is a pattern? Describe some patterns you may have seen around our room. What patterns are found outside our classroom (pinecone, fences, pavement).

EXPLAIN — From the calendar on a wall to the ceiling tiles, patterns are found all around us. Some patterns cover a whole space by mixing and matching shapes. Demonstrate patterns from polygons using a great online interactive "Tesselation Creator."

http://illuminations.nctm.org/Activity.aspx?id=3533

MOVE — Take a walk, and look for patterns. Challenge students to locate at least 10 patterns around the school.

INVESTIGATE

Practice with tesselation patterns. Print a pattern on cardstock. Kids trace the pattern anywhere on a sheet of paper. Fit the shapes against each other to cover the whole page. Decorate so that no shapes that touch have the same color.

Challenge kids to create a pattern that covers an entire sheet of paper. Stick to squares and rectangles to make the exercise simpler. Incorporate your own lessons on color, size and attributes.

ELA — Make a word wall with color, size and texture focus.

NEW HOME FOR PET
STEMVESTIGATION: SIZE AND SCALE

MATHEMATICS

READ   How Big Is a Guinea Pig *page 92*

DISCUSS   Animals come in all shapes and sizes. Let's name some dogs that fit these words; tiny, small, average, big, large, and giant.
Tiny = Teacup Poodle   Small = Jack Russell   Average = Golden Retriever
Big = German Shepherd   Large = Mastiff or Sheepdog   Giant = Great Dane

EXPLAIN   Sometimes agreeing on what sizes fit in a category is tricky. An animal one person considers tiny might just be small for someone else. Create a system of measurement so that all descriptions agree.

MOVE   Call out a descriptor. Students have 10 seconds to locate and stand next to an item that matches the size. For example, call out, "small." Students stand next to their "small" items. Each student must name and defend choices. Play in teams or as a whole class.

INVESTIGATE

Small groups work together to devise a set of measurements that fits each adjective for size. Use measurements, pictures, or animals as the standard. Create a chart from drawings or cut-and-paste images.

Size Word	From	To	or
tiny	ant	hamster	1/2 inch to 5 inches
small	guinea pig	rabbit	51/2 in to 1 foot
average	cat	boxer dog	13 inches to etc....
big	bobcat	deer	**or**
large	great dane	pony	80 lbs to 200 lbs
giant	horse	elephant	201 lbs to 1 ton

ELA   Use similes to describe pets. Ex: The dog is as big as a chair.

NEW HOME FOR PET
PLAY WITH SHAPES

HOOP AND HOLLER

MATERIALS
hula hoops

SETUP
Place a hula hoop per team on the ground about 30 feet from the kids.

INSTRUCT THE PLAYERS
You'll need one hoop per team.
Divide the class into several small teams.
The first player of each team runs to their team's hula hoop
which is placed about 30 feet in front of each team.

The team does five jumps in and out of the hoop
and returns to the group.
The first team to finish the relay wins.

INSTRUCT THE PLAYERS
Form teams of 4.
Line your team up in front of a hula hoop (or jump rope).

NEW HOME FOR PET

RHOMBI'S ADVENTURES IN 3D
UNIT 3: HABITATS & BASIC NEEDS

NEW HOME FOR PET

Once upon a time there was Rhombi. She had a sturdy home shaped like a cube. A pyramid roof protected her house from the sun, wind, and rain. Sunny spring days had finally arrived.

NEW HOME FOR PET

Rhombi took a walk. She saw all kinds of pets with their people. There were puppies. There was a bunny. She saw a guy walking with a lizard on his shoulder. Two kids with their cat made up her mind. She wanted a kitten.

NEW HOME FOR PET

March 14

March 15

She picked a scruffy little cat. Pet got a bath, a new collar, and a pillow on Rhombi's bed. Each night the tiny kitten curled up right under Rhombi's chin. Things were great while Pet was little. As she grew, the cat needed more room to run and play.

NEW HOME FOR PET

Remember Pet's veterinary appointment on May 1st. She's needs vaccinations!

Aunt B's Pet Center
3.14 Pi Avenue
GeoCity, USA

TO: Rhombi
011235813 Fibonnaci St
Geocity, USA 213456

May 1

In the first week of May, Rhombi found Pet in the sink. "What are you doing, Pet," she laughed and scratched the kitty under her chin. In the second week of May, Rhombi found Pet in the cabinet. Pet was a little sulky and would not play.

NEW HOME FOR PET

Rhombi started to worry about the little cat. Pet was such a happy kitten, Rhombi wondered what was making her so grumpy? In the third week of May, Rhombi hopped to stay on her feet when Pet scooted across the room. Pet grumbled a little.

NEW HOME FOR PET

May 29

In the fourth and last week of May, Rhombi tripped over Pet three times. By the first week in June, Rhombi and Pet were grumpy about the tiny space. That week, Rhombi made a decision. For Pet's sake, she needed to give the kitty more space to play.

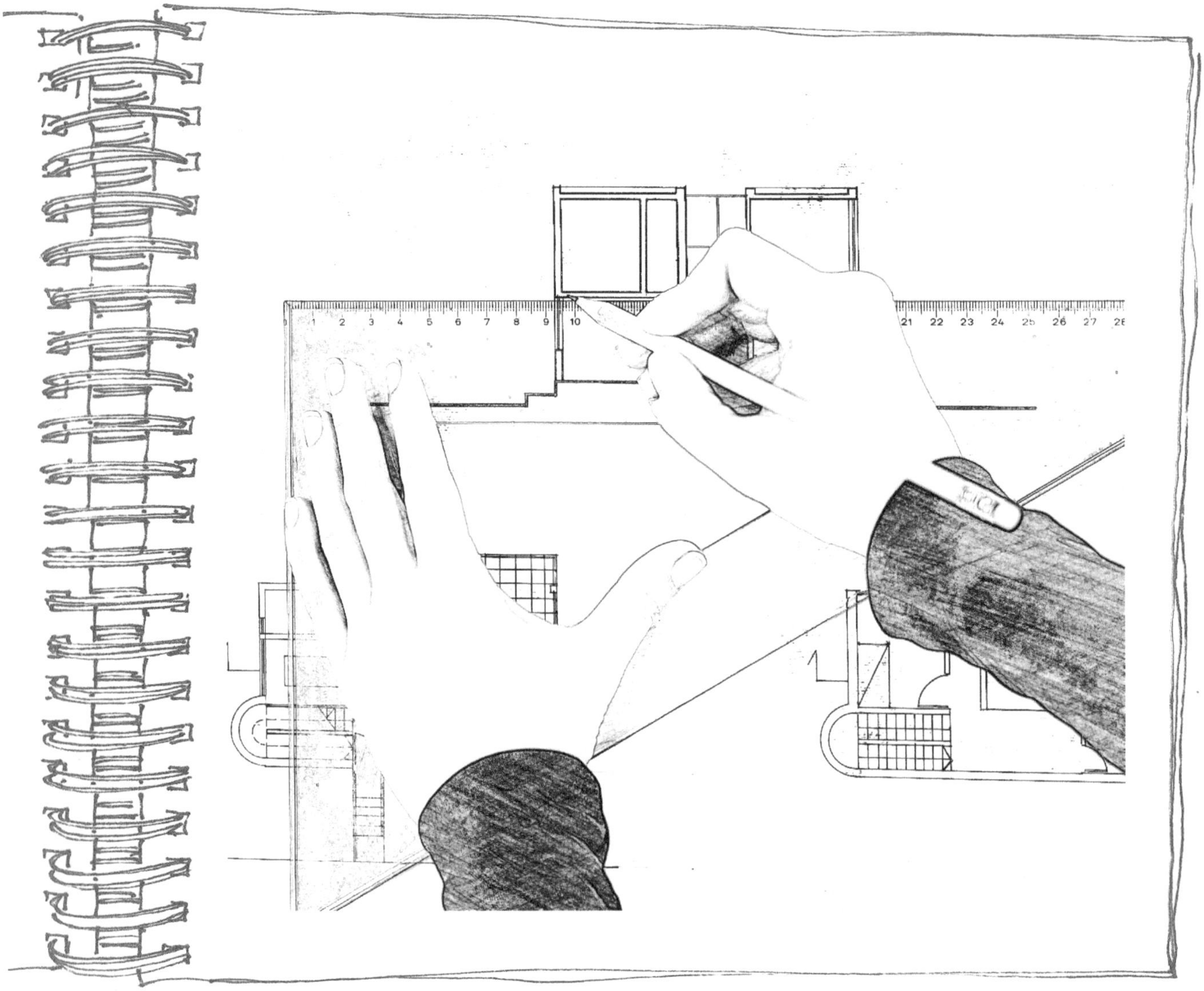

NEW HOME FOR PET

Rhombi realized that she needed a special place for her new pet. She looked out beyond the pages of her world and asked, "Designers, can you help me create a small house for Pet?"

NEW HOME FOR PET
TALK ABOUT S.T.E.A.M.

How many square faces does a pyramid have? **1**

How many triangular faces does a pyramid have? **4**

What is the total number of faces on a pyramid? **5**

Make a list of all the pets you can remember. **Answers will vary.**

Use the list to make a tally sheet counting students' pets.

Turn the tally sheet into a class bar graph.

What kind of animal did Rhombi choose? **kitten**

Why would Rhombi keep Pet inside? **Pet would get too cold outside.**

What food would Rhombi eat for breakfast? What does Pet eat? **Answers will vary.**

Make a list of breakfast foods and tally food eaten by students.

What animals have a tail? **Answers will vary.**

What sounds would a grumbling pet make? **Ex: whine, chirp or meow, etc.**

Why might Pet scoot across the room? **She might be chasing toys.**

What area of floor would be "tiny" for you? Measure and test. **Ex: A 4ft x 4ft room is tiny.**

What do you think Rhombi has decided? **She will need a room for pet.**

Why does Rhombi need a special room for Pet? **She and Pet need more room to move.**

What environment is needed for a puppy? a cat? a snake? a bird? a frog? a butterfly?

What materials will you need in order to build Rhombi's pet's house? **squares, triangles, tape, and other materials to make Pet comfortable...**

RHOMBI'S ADVENTURES IN 3D
UNIT 3 CHALLENGE: NEW HOME FOR PET

"Designers, can you help me build a room for Pet?"
The room must use triangles and squares to make a 3 dimensional room.

Draw Rhombi's playhouse with its pyramid roof.	Draw Rhombi's playhouse with a new room just for Pet!

Draw each of the pieces you will need to complete the challenge.

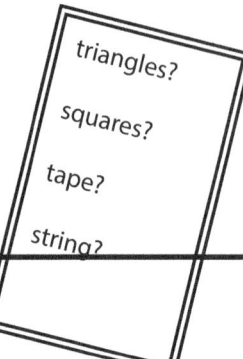

triangles?
squares?
tape?
string?

On the back of this paper, write out the steps you will need to take.

RHOMBI'S ADVENTURES IN 3D
UNIT 3 RUBRIC: NEW HOME FOR PET

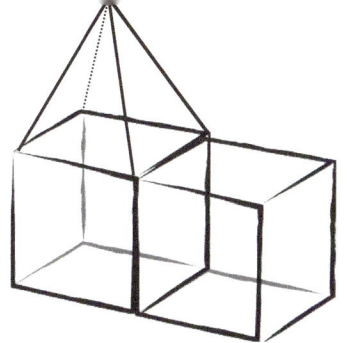

TASK: Students will use their knowledge of habitat, house, pyramids and strength to build a scale model of a habitat for a Pet.

	Content	Organization	Design & Build
1	• Is well thought out and supports the solution to the challenge or question • Reflects application of critical thinking • Has clear goal and reasons for designing this habitat for the pet • Is accurate	• Information is clearly focused in an organized and thoughtful manner • Information is constructed in a logical pattern to support the solution	• Uses 2D shapes to create 3D shapes • Neatly joins corners and edges • Constructs habitat that reflects the needs of a chosen pet
2	• Supports the solution • Has application of critical thinking that is apparent • Has no clear goal • Has some factual errors or inconsistencies (hamsters do not live in fish bowls full of water) • Student is unable to adjust concept when original idea is questions	• Project has a focus but might stray from it at times (more concerned with form than function) • Information appears to have a pattern, but the pattern is not consistently carried out in the project	• Identifies a square but does not combine them to make a cube • Edge seams are mis-joined and uneven • Constructs habitat that does not reflects the needs of a chosen pet
3	• Provides inconsistent information for solution • Has no apparent application of critical thinking • Has no clear goal • Has significant factual errors, misconceptions, or misinterpretations	• Content is unfocused and haphazard • Information does not support the solution to the challenge or question • Information has no apparent pattern	• Does not draw/make 2D shapes • Does not know to join squares to make a cube • Edge seams are missing or ragged • Student project falls apart and student is unable to make changes

PET'S CELEBRATION

RHOMBI'S ADVENTURES IN 3D
UNIT 4: STRUCTURES

RHOMBI'S ADVENTURES IN 3D
UNIT 4: PET'S CELEBRATION

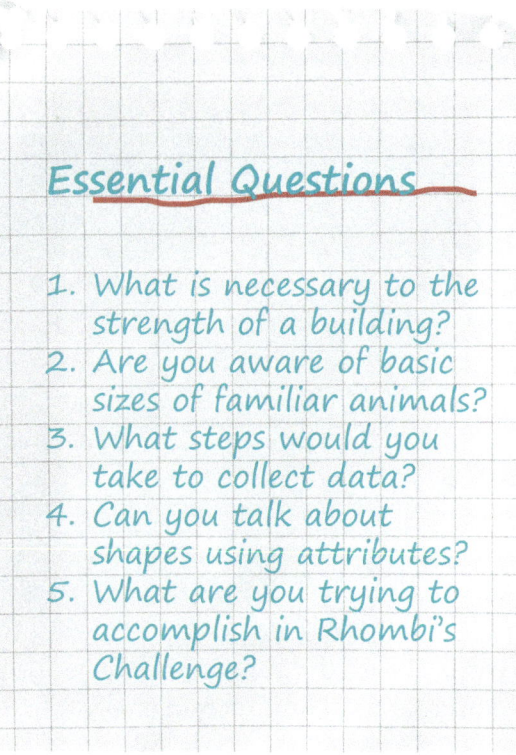

Essential Questions

1. What is necessary to the strength of a building?
2. Are you aware of basic sizes of familiar animals?
3. What steps would you take to collect data?
4. Can you talk about shapes using attributes?
5. What are you trying to accomplish in Rhombi's Challenge?

ENDURING UNDERSTANDING

Two dimensional shapes create new shapes.

"Halves" means two equal parts."

Circles and rectangles break into equal parts.

Breaking circles or rectangles into more equal parts means that the parts will be smaller.

Three objects may be put in order from longest to shortest by comparing their lengths.

The length of an object may be discussed using whole numbers.

A smaller object may be used as a measurement tool when measuring longer lengths.

Larger numbers consist of tens and ones.

Organizing information helps in discussion.

PET'S CELEBRATION

Focus on combining two-dimensional shapes to create three-dimensional shapes for a purpose. Investigate structures and material strength. Discuss celebrations. Investigate materials, length, time, and weight. Create mosaics using squares and triangles.

Rhombi celebrates Pet's birthday with a party for Pet's closest friends. Without enough space, Rhombi asks her designer friends to help wiht a celebration space.

autumn
the season between summer and winter; in the Northern Hemisphere, autumn lasts from the September equinox to the December solstice

column
vertical pillar used to support a structure

vertices
intersection where two planes meet

primary colors
In dyes, pigments, and paints, the primary colors are red, yellow, and blue

PET'S CELEBRATION
CURRICULUM CONNECTIONS

Rhombi's Pet is having a party and needs a space to celebrate.

ART CONNECTIONS - Mosaics

According to WiseGeek.org, mosaic is an art form that involves fitting small pieces of rock, shell, tile, or glass called tesserae together to create a pattern that may be abstract or representational. Students at Eisenhower Junior High School in Taylorsville, Utah hold the World Record for Largest Post-It Mosaic (38,400 pieces of paper).

Visit HistoryWorld.net for examples of mosaics.

TECHNOLOGY CONNECTIONS

Visit QRStuff to generate codes for use in the classroom. To share the code, paste it into a Word Doc or print directly from the page.

HISTORICAL CONNECTIONS

"A column is a vertical pillar that is used to support the structure of a building. In classic architecture, a column rests on a base and is mounted by a lid, called a capital." The Greeks used columns to build open-air structures.

The three styles of Greek column are Doric, Ionic, and Corinthian. Examples of Greek columns may be seen in the following buildings:
Rome, Italy's Pantheon and Colosseum;
Egypt's Temple of Amon
and USA's Lincoln Memorial.

MATH BACKGROUND - Yard Sticks

"A GOOD unit of measurement," writes Robert Crease, "must satisfy three conditions." It has to be easy to relate to, match the things it is meant to measure in scale (no point using inches to describe geographical distances) and be stable. The yardstick did not always meet those criteria.

Its history is sketchy, but most accounts agree that the yardstick was coined by King Edward I (reigned 1272-1307) who declared, "three feets make one yard." The yard was based on the length of a pendulum that took one second to complete its swing.

MINDBUGS TO NOTE

ASSESS NUMERACY
Hold several objects in your right hand.
Show the student.
Assess how far along the spectrum your student has progressed:
- Physically touches each item.
- Mentally touches each item.
- Recognizes the number of objects.

Physical contact required: issue manipulatives for all counting and problem solving.
Mental contact required: ready for activity books and handouts.
Recognition: ready for addition and subtraction.

*Test all students, including those identified as accelerated. You may be surprised at the outcomes.

PET'S CELEBRATION
CURRICULUM CONNECTIONS

COMMON CORE CONNECTIONS

ELA/Literacy
RI.K.1 With prompting and support, ask and answer questions about key details in a text. **W.K.1** Use a combination of drawing, dictating, and writing to compose opinion pieces in which they tell a reader the topic or the name of the book they are writing about and state an opinion or preference about the topic or book. **W.K.2** Use a combination of drawing, dictating, and writing to compose informative/explanatory texts in which they name what they are writing about and supply some information about the topic.

Mathematics
MP.2 Reason abstractly and quantitatively. **MP.4** Model with mathematics. **K.CC.A** Know number names and the count sequence. **K.MD.A.1** Describe measurable attributes of objects, such as length or weight. Describe several measurable attributes of a single object. **MD.A.1** Describe measurable attributes of objects, such as length or weight. Describe several measurable attributes of a single object.

NEXT GEN SCI CONNECTIONS

ETS1.A: Defining Engineering Problems
A situation that people want to change or create can be approached as a problem to be solved through engineering. Such problems may have many acceptable solutions.

1-ESS1-1,1-ESS1-2 Patterns in the natural world can be observed, used to describe phenomena, and used as evidence.

ESS2.D: Weather and Climate
Weather is the combination of sunlight, wind, snow or rain, and temperature in a particular region at a particular time. People measure these conditions to describe and record the weather and to notice patterns over time. (K-ESS2-1)

K-ESS2-1. Use and share observations of local weather conditions to describe patterns over time.

UNIT 4 S.T.E.A.M. ACTIVITIES FOR PET'S CELEBRATION

Rhombi Audio Download / Video (available December 2014)

UNIT 4: Pet's Celebration:	Science:	Modify desserts safe for Pets and people.
UNIT 4: Pet's Celebration:	Technology:	Explore mass and weight with digital scales.
UNIT 4: Pet's Celebration:	Engineering:	Investigate the strength of columns.
UNIT 4: Pet's Celebration:	Art & Music:	Create mosaics from squares and triangles.
UNIT 4: Pet's Celebration:	Mathematics:	Measure to build a sense of size and scale.
UNIT 4: Pet's Celebration:	ELA:	Read and illustrate "PET'S CELEBRATION."

PET'S CELEBRATION
BUILD S.T.E.A.M. WITH GREAT BOOKS

SCIENCE
Butterfly Birthday
Harriet Ziefert

Classroom Library
Pets
Nutrition

TECHNOLOGY
Who Made This Cake?
by Chihiro Nakagawa

Classroom Library
Measurement
Scales
Construction
Weight and Mass

ENGINEERING
If You Lived Here: Houses of the World
Giles Laroche

Classroom Library
BridgeS
Travel
Animal Migration

ART & MUSIC
Birthdays Around the World Library
by Mary D. Lankford

Classroom Library
Mosaics
Eric Carle Books

MATH
The Big Birthday Surprise: Life Lessons with Junior
Dave Ramsey

Classroom Library
Measurement, How String Is Made

PET'S CELEBRATION
STEMVESTIGATION: DESSERT

READ   **HONEY, HONEY** *page 96*

DISCUSS   When is your birthday? Let's mark the calendar. How do you celebrate? Do you have any traditions? What do you eat? Do you let your pets celebrate with you? Do they eat the same foods as people?

EXPLAIN   Dogs cannot eat chocolate, and cats cannot drink milk. Most pets get sick when eating avocado, including dogs, cats, horses, birds and rodents. Also skip the nuts, raisins and plums. So, what would you serve to pets that attend your party?

MOVE   Mash. Mash. Mash to make some homemade treats!

INVESTIGATE

Prepare for the activity by slicing at least 2 bananas (and any other fruit you want to include) per student and freezing. To cut chaos, go ahead and freeze the fruit in individual baggies. Double bag for safety's sake. Provide cups, saucers, cardboard pieces, and other cast-off materials for constructing the "smooshers."

You are challenged to make a smooshy mooshy "ice dreamy" treat safe for pets to eat. Using materials provided, create a food "smoosher" that will mash the frozen fruit before it thaws. You'll end up with a creamy healthy treat that's safe for everyone (unless you are allergic to bananas).

When done, taste test the delicious snack!

ELA   Create directions for the next class to make smooshers.

PET'S CELEBRATION
STEMVESTIGATION: DIGITAL TOOLS

TECHNOLOGY

READ Waxing Colorful *page 98*

DISCUSS Have you ever seen a parent or caregiver use a scale to weigh themselves or to weigh food?

EXPLAIN A scale measures the pull of objects to the planet based on gravity. When an object is placed on the tray, a flexible (bendy) piece inside the tool bends. A little piece of foil is pushed to change the flow of electrical current. A signal goes to the screen showing the object's weight on Earth. Basically, pressing the top sends a signal. More pressing means more weight.

MOVE Locate 5 objects to weigh on the scale. (This helps them begin to estimate the weight of objects less than 5 pounds.)

INVESTIGATE

Use the MindBugs Digital Scale activities to weigh items around the classroom. Upon introduction to the tool, explain its use including all the technical terminology above. Also, show students how the scale works. Set it for ounces or grams as desired before having students work with the device. When they write down answers, students should absolutely write down the decimal and parts of a whole. Introducing the vocabulary now will help understanding later. Educators use extensive vocabulary in Language Arts, and schools should begin the same habit for STEAM subjects.

ELA Make a chart to show object and weight for 5 items.

PET'S CELEBRATION
STEMVESTIGATION: STRENGTH

ENGINEERING

READ	**Strength in Numbers** *page 100*
DISCUSS	What will happen if I step on this tube? Elicit responses, then step evenly on the top of the tube. They may observe on their own that a slight lean in any direction causes the tube to crumple more quickly on that side. Placing a piece of cardboard over the top of the tube may help distribute weight more evenly.
EXPLAIN	Some structures are supported by columns. Patios, porches, basements, and bridges are places you may see columns. The ones you cannot see are hidden in walls. The outer walls are not strong enough to hold up a building's entire weight. Other materials fill the column to make it stronger.
MOVE	Students try balancing items on their own tubes (paper towel or toilet paper) to test strength.

INVESTIGATE

Given construction paper or cardstock, challenge teams to devise a column that -when stood vertically- holds 1 pound of weight. There are many possible solutions:
- Increase the stiffness of the sides of the tube to help the structure resist buckling under a load.
- Decrease the possibility of collapse by filling the tube. The load is distributed evenly by the material inside the tube. A column can be filled with cheaper material and still increase the column's compression strength.

Test the strength of columns with a 1-pound book. Chart tube thickness vs weight held. Chart material in the column vs weight held.

PET'S CELEBRATION
STEMVESTIGATION: GEOMETRY IN MOSAICS

The ARTS

READ — Square in the Middle *page 102*

DISCUSS — Have you ever seen a mosaic? Has the art class ever created mosaics before? What do you think a mosaic might be?

EXPLAIN — According to WiseGeek.org, mosaic is an art form that involves fitting small pieces of rock, shell, tile, or glass called tesserae together to create a pattern that may be abstract or representational. In Rhombi's words, fit little shapes together to make a big shape.

MOVE — Create a people mosaic. Lie down to make a shape (flower or sun is easiest), and make sure to take a photo!

INVESTIGATE

Cut colored paper into one inch strips. Students will use age-appropriate scissors to cut strips down to small squares for use in the mosaic.

Students draw a very simple picture in white crayon on black construction paper. Then, working from the center to the outside, cover the image with colored squares. For each "tile" used, touch the tile to a glue-saturated sponge to minimize mess. Color and placement are all child's choice. Let them explore.

ELA — Use your mosaic to tell a picture story.

PET'S CELEBRATION
STEMVESTIGATION: LENGTH

READ   **Comin' Up a Cloud** page 104

DISCUSS   Are you taller or shorter than this stick? What might be wider than this stick?

EXPLAIN   This stick is a yardstick that has 3 big sections and 36 little sections. The 3 are called feet. The 36 are called inches. Today we'll use the yardstick to measure our string. Each piece of string will be one yard, or 3 feet or 36 inches long.

MOVE   Have everyone mark their heights on the whiteboard. and compare to the yardstick.

INVESTIGATE

Place string along the length of the yardstick. Cut string to the same length as the stick. Using the string and glue, create a shape picture. You'll know that your finished shape measures exactly one yard, 3 feet or 36 inches!

ELA   Students explain the process of creating their present their shapes to older students.

PET'S CELEBRATION
PLAY WITH SHAPES

HOPSCOTCH

MATERIALS
chalk
surface safe for drawing and jumping
cut sponges (or rocks if you are feeling particularly bold)

SETUP
Draw hopscotch boards for young children.
Encourage older students to draw their own game boards.

INSTRUCT THE PLAYERS
The first player tosses a marker (rock, coin) into the first square; it must land within the confines of the square without bouncing out or touching a line.
The player then hops through the course, making sure to skip the square with the marker in it. Players hop in single squares with one foot (either foot is fine), and use two feet for the side by side squares, one in each square. Upon completion of the hop sequence, the player continues her turn, tossing the marker into square number two and repeating the pattern. Players begin their next turn where they last left off. Player loses a turn (or ends the current turn) if these things happen:
- player steps on a line;
- player misses a square on toss;
- or the player loses balance.

INSTRUCT THE PLAYERS
Explain the rules.
Demonstrate one full turn.
Have students announce each number as it is hopped, counting up to 10 and back down to 1.

PET'S CELEBRATION

RHOMBI'S ADVENTURES IN 3D
UNIT 4: STRUCTURES

PET'S CELEBRATION

Once upon a time there was Rhombi. Rhombi lived in a house shaped like a cube with a cool pyramid roof. Her house had 8 edges that she painted a different color each year. The house would not be toppled by Autumn winds.

PET'S CELEBRATION

The pointed pyramid roof kept out Winter rain and snow. Rhombi's pet had a place of its own with 8 corners (vertices) where Pet stored her treasures. Pet's place stayed dry in the Spring rain and cool in the Summer sun.

PET'S CELEBRATION

Rhombi was planning Pet's Summer birthday party and wanted to invite many forest friends. Rhombi and Pet printed circus party invitations on primary colored paper. Pet wanted 10 friends to attend her party.

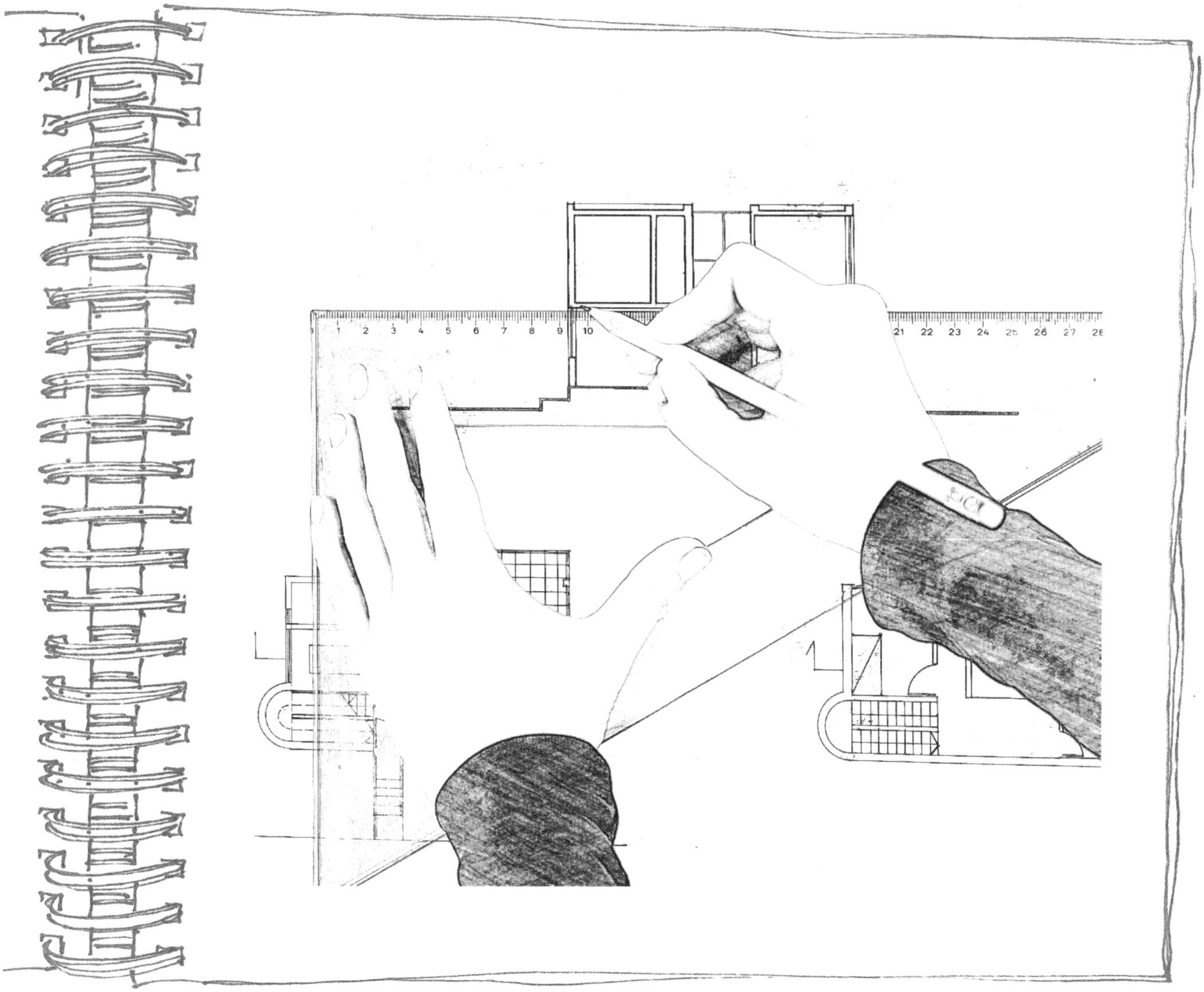

PET'S CELEBRATION

Rhombi realized that she didn't have enough room for a house full of friends. She decided to ask for help. She looked beyond the pages of her world and asked, "Builders, can you help me make a space for friends?"

PET'S CELEBRATION
TALK ABOUT S.T.E.A.M.

What 3D shapes are part of Rhombi's house? **cube, pyramid**

How many edges does a pyramid have? **8 edges**

Rhombi lived through what two seasons in her new house? **Fall and Winter**

How many vertices does Pet's new room have? **8**

Graph all the birthdays by month. Graph all the birthdays by day.

Plan a party! What will you eat? How will you decorate?

If you have a party budget of $20, how will you spend the money?
Consider using catalogs or ads to actually plan the menu and budget.

What 3 colors are described as primary? **red, blue, yellow**

What types of animal friends might live in the forest?
birds, rabbits, foxes, turtles (any appropriate animals or insects)...

How much space will Rhombi need to add to her home?
multiply Pet's space by ten

Will you design a party porch? Will you design a whole new room?

What materials will you need in order to build Rhombi's pet's house?
Answers will vary.

RHOMBI'S ADVENTURES IN 3D
UNIT 3 CHALLENGE: PET'S CELEBRATION

"Designers, will you create a place for Pet's party?"
The room must use triangles and squares to make a 3 dimensional room.

Draw Pet's house.	Draw Pet's celebration space.

Draw each of the pieces you will need to complete the challenge.

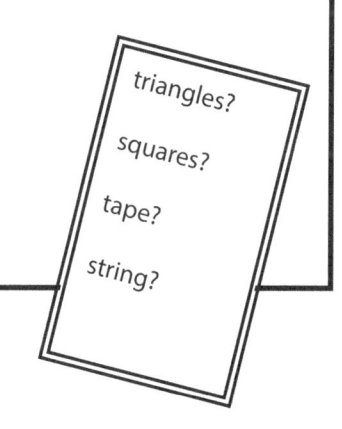

triangles?

squares?

tape?

string?

On the back of this paper, write out the steps you will need to take.

RHOMBI'S ADVENTURES IN 3D
UNIT 4 RUBRIC: PET'S CELEBRATION

TASK — Students will use their knowledge of habitat, house, pyramids and strength to build a scale model of a celebration space for pets.

	Content	Organization	Design & Build
1	• Is well thought out and supports the solution to the challenge or question • Reflects application of critical thinking • Has clear goal that is related to the topic • Is accurate	• Information is clearly focused in an organized and thoughtful manner • Information is constructed in a logical pattern to support the solution	• Uses 2D shapes to create 3D shapes • Neatly joins corners and edges • Constructs a platform with strong pillars able to withstand given weight
2	• Supports the solution • Has application of critical thinking that is apparent • Has no clear goal • Has some factual errors or inconsistencies (i.e. wheels do not rotate around axle)	• Project has a focus but might stray from it at times (more concerned with form than function) • Information appears to have a pattern, but the pattern is not consistently carried out in the project	• Edge seams are mis-joined and uneven • Constructs an adequate platform that does not support weight
3	• Provides inconsistent information for solution • Has no apparent application of critical thinking • Has no clear goal • Has significant factual errors, misconceptions, or misinterpretations	• Content is unfocused and haphazard • Information does not support the solution to the challenge or question • Information has no apparent pattern	• Edge seams are missing or ragged • Student project falls apart and student is unable to make changes

RHOMBI'S GALLERY

**RHOMBI'S ADVENTURES IN 3D
UNIT EXTENSION**

RHOMBI'S GALLERY
CURRICULUM CONNECTIONS

Rhombi is opening a business, and she needs help with the building.

ART CONNECTIONS - Logo

A logo is a piece of art that uses graphics to convey a message or help consumers recognize a product. The logo is a design that includes a name & symbol or acronym identifying and/or distinguishing your brand from others. A logo uses a SYMBOL (emblem, icon, sign) in its DESIGN (the "look and feel" including lettering, color, shape, idea, pattern).

TECHNOLOGY CONNECTION

Clipart Collections
http://school.discoveryeducation.com/clipart/

HISTORICAL CONNECTIONS

Some of the world's current businesses have stood the test of time. Lloyd's of London started life in 1688 as Edward Lloyd's Coffee House. Japan's Kongo Gumi temple builders had a 1,428 year run that ended in 2006.

The first official logo was trademarked in 1876 by Bass Ale, but individuals and companies have used personal "seals" for thousands of years. A carved stamp was pressed into hot wax so recipients would know that the origin of the document was authentic.

MATH BACKGROUND - Number Line

As explained by www.k-5mathteachingresources.com, the number line was originally proposed as a model for addition and subtraction by researchers from the Netherlands in the 1980s. A number line is a visual representation for recording and sharing students' thinking strategies during the process of mental computation. Students will begin to solve problems mentally by picturing the number line in their heads. A number line is a line in which real numbers can be placed, according to their value. Each point on a number line corresponds to a real number. Each real number has a unique point that corresponds to it. Ex: the number 2.5 (2 1/2) corresponds with the point on a numberline that is halfway between two and three.

MINDBUGS TO NOTE

- Students confuse the minute and hour hands.
- Students have difficulty estimating the duration of a given length of time.
- Digital clocks and timers have number scales based on 60 not 100.
- Students often forget to begin measuring a length from the zero mark.
- They use the edge of the ruler or start at 1.
- When measuring lengths longer than the ruler, some students flip the ruler over and over.

RHOMBI'S GALLERY
CURRICULUM CONNECTIONS

COMMON CORE CONNECTIONS

ELA/Literacy
RI.K.1 With prompting and support, ask and answer questions about key details in a text. **W.K.1** Use a combination of drawing, dictating, and writing to compose opinion pieces in which they tell a reader the topic or the name of the book they are writing about and state an opinion or preference about the topic or book. **W.K.2** Use a combination of drawing, dictating, and writing to compose informative/explanatory texts in which they name what they are writing about and supply some information about the topic.

Mathematics
MP.2 Reason abstractly and quantitatively. **MP.4** Model with mathematics. **K.CC.A** Know number names and the count sequence. **K.MD.A.1** Describe measurable attributes of objects, such as length or weight. Describe several measurable attributes of a single object. **MD.A.1** Describe measurable attributes of objects, such as length or weight. Describe several measurable attributes of a single object.

NEXT GEN SCI CONNECTIONS

ETS1.A: Defining Engineering Problems
A situation that people want to change or create can be approached as a problem to be solved through engineering. Such problems may have many acceptable solutions.

1-ESS1-1, 1-ESS1-2 Patterns in the natural world can be observed, used to describe phenomena, and used as evidence.

ESS2.D: Weather and Climate
Weather is the combination of sunlight, wind, snow or rain, and temperature in a particular region at a particular time. People measure these conditions to describe and record the weather and to notice patterns over time. (K-ESS2-1)

K-ESS2-1. Use and share observations of local weather conditions to describe patterns over time.

EXTENSION S.T.E.A.M. ACTIVITIES FOR RHOMBI'S GALLERY

EXTENSION Audio Download / Video (available December 2014)

EXTENSION:	Rhombi's Gallery:	Science:	Explore options for waterproofing materials.
EXTENSION:	Rhombi's Gallery:	Technology:	Use the stopwatch to talk about elapsed time.
EXTENSION:	Rhombi's Gallery:	Engineering:	Take virtual tour of skyscrapers.
EXTENSION:	Rhombi's Gallery:	Art & Music:	Create a logo that reflects your personality.
EXTENSION:	Rhombi's Gallery:	Mathematics:	Develop a day's timeline.
EXTENSION:	Rhombi's Gallery:	ELA:	Read and illustrate "Rhombi's Gallery."

RHOMBI'S GALLERY
BUILD S.T.E.A.M. WITH GREAT BOOKS

SCIENCE

Butterfly Birthday
Harriet Ziefert

Classroom Library
colonial days
weather
rain
snow

ENGINEERING

If You Lived Here: Houses of the World
Giles Laroche

Classroom Library
towers
towns
architecture

MATH

Classroom Library
historical biographies
timelines
number lines
calendars

TECHNOLOGY

Who Made This Cake?
by Chihiro Nakagawa

Children's Video Tour for The National Gallery of Art
http://www.nga.gov/education/timetravel/

Classroom Library
time
telling time
clocks
biographies of inventors

ART & MUSIC

Birthdays Around the World Library
by Mary D. Lankford

Classroom Library
cartooning
careers
starting a business
drawing
symbols

RHOMBI'S GALLERY
STEMVESTIGATION: WATERPROOF

TOUR — The National Gallery of Art Video Tour for Kids
http://www.nga.gov/education/timetravel/

DISCUSS — What is waterproof?
Do you have any items that are waterproof?

EXPLAIN — Waterproof materials are designed to keep liquid from entering or passing through the material. People used beeswax, lanolin from sheepskin, flaxseed, linseed oil and other natural materials before chemical ones were invented. Paper can be waterproofed using melted crayon wax, tape, wax paper, plastic bags or aluminum foil.

MOVE — Create a large chalk outline. Students spread evenly through the area. Change the outline. Repeat.

INVESTIGATE

Challenge students to "waterproof" a piece of paper using materials in the classroom scrap box. Provide plenty of test paper and various "waterproofing" options. Students use a dropper or tablespoon to maintain a consistent amount of liquid poured on the papers. Place test papers in a tray to minimize mess.

Students create a chart, table or graph to show the item and the number of seconds or minutes that pass before the liquid saturates the paper. **Provide materials.** plastic baggies, petroleum jelly, wax paper, beeswax, old candles to rub on paper, foil, old broken crayons make a great waxy water-repellent layer... color thick layers, then leave in sunlight or under a strong heat lamp to melt...

ELA — Each hour of the day, take digital photos to aid sharing.

RHOMBI'S GALLERY
STEMVESTIGATION: ELAPSED TIME

READ   Mindbugs in Time: Elapsed Time

DISCUSS   How many students have participated in sports? Did you see a referee using a stopwatch? What about spelling bees or other activities that must be timed.

EXPLAIN   Stopwatches measure elapsed time, the number of seconds that pass from start to finish of an activity. (Demonstrate the stopwatch.)

MOVE   How many jumping jacks can you do in a minute? How many seconds pass while you do 10 lunges?

INVESTIGATE

Use the stopwatch while lining up for breaks, during bathroom trips, lunchroom cleanup, etc. Choose a set of activities from the Mindbugs Stopwatch Activity Guide. Challenge students to master the stopwatch while timing everything from jumping jacks to repeating the alphabet.

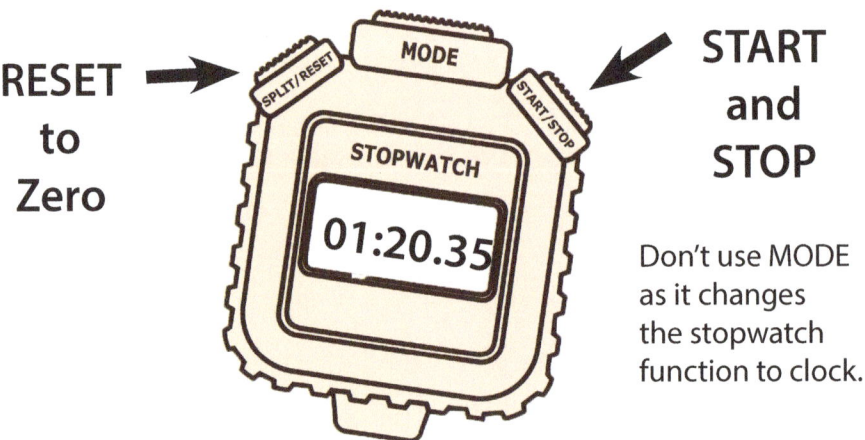

ELA   Demonstrate the stopwatch to another student!

TECHNOLOGY

RHOMBI'S GALLERY
STEMVESTIGATION: TALL TOWERS

ENGINEERING

WATCH   **PBS Video Tour of Skyscrapers**
http://video.pbs.org/program/super-skyscrapers/

DISCUSS   How many people have seen tall towers like the ones in a city or silos in the country? How many layers (floors) do really tall towers have?

EXPLAIN   Engineers have things to overcome like goals, money, limited materials, and a time limit. An engineering team that can design a structure to meet the objectives with the fewest materials (hence, less cost), is more likely to get the job than teams that cannot handle the task.

MOVE   Collect newspapers!

INVESTIGATE

Challenge students to design and construct a model tower using only newspaper and tape and scissors. Each team has limited supplies and must finish in one class block. Make the tower as tall as possible but able stand up to a hair dryer held 4 feet away on low speed.

As needed, students may be shown the paper roll option. Roll newspaper and tape to create a tube. Fit tubes together to make a tower.

Students record tower height and try to beat their own records.

ELA   Describe the process of making your tower well enough that someone else can repeat the activity and get the same outcome.

RHOMBI'S GALLERY
STEMVESTIGATION: LOGOS & GRAPHICS

The ARTS

VISIT   View 20 Popular Kids Logos
http://www.findthatlogo.com/20-popular-kids-logos-brands

DISCUSS   Have you noticed the pictures that companies use to get your attention? Can anyone name some logos that you see on your clothing?

EXPLAIN   These logos are messages to buyers. Businesses want to talk you into buying their products or using their services. Logos may be words that look like a picture called a font. Logos may be pictures like animals or toys. All logos represent the owner in some way.

MOVE   Take a logo walk through the school or classroom to locate various examples.

INVESTIGATE

Challenge students to create a logo for their own business. The logo should show a "symbol" like animal or toy or tool. The logo should have the student's name or initial's as in "Tony's Toy Company."

1. What are your favorite businesses to visit or activities to attend? What do you like to do for fun?
   ex: I like to build block towers. My favorite place is the playground.

2. How could you provide that service for other people?
   ex: I could make cool towers for playgrounds

3. What is a picture that would help people know what you can do for them? Where will you put your name?
   ex: a block tower with my name on the blocks

ELA   Make an advertisement for the business.

RHOMBI'S GALLERY
STEMVESTIGATION: TIME AND SCHEDULES

READ Visit an interactive **Black History Timeline** at Cocoa's Kids Read   http://www.cocoakidsread.com/blackhistory/

DISCUSS Do you remember what you were doing this morning? Can you recall what you ate for breakfast? If I ask you to remember what you did after school last Thursday, can you describe the activity?

EXPLAIN People create timelines to see the flow of events. These lines help people visualize the where and when in history.

MOVE March along a schoolyard timeline. Place sticks or draw chalk lines along a path. Write the current year. Walk back to the year most students were born. Using chalk, help them fill in the years with school events and relative ages when children begin walking, etc.

INVESTIGATE

Using the handout as necessary, challenge students to create a timeline for the week. Each day from Monday through Thursday do the following:
- Draw or describe a memorable AM event.
- Draw or describe lunch.
- Draw or describe a PM event.

Great online timeline tools for teachers link to web resource.
EDUCATIONAL TECHNOLOGY AND MOBILE LEARNING website   http://alturl.com/oyrvr
ARTISTS HELPING CHILDREN website   http://alturl.com/t98i5

Events may not match for every student. The goal is to generate unique timelines for Friday's discussion.

ELA On Friday, share timelines with the class.

RHOMBI'S GALLERY
PLAY WITH SHAPES

CLIP ARTISTS

MATERIALS
whiteboard and whiteboard markers
or
markers and large paper

SETUP
Use geometry words and vocabulary from all subjects or attribute cards you already own to fill a bag with options.

INSTRUCT THE PLAYERS
Break the class into two teams. Team one draws a word to be illustrated. One player draws while others guess the word.

Team two must keep the same word and find a different way to represent its meaning. THEN, pull a new word to illustrate.

Play ends after 20 minutes.

RHOMBI'S GALLERY

**RHOMBI'S ADVENTURES IN 3D
UNIT EXTENSION**

RHOMBI'S GALLERY

Once upon a time, there was Rhombi. Rhombi had a wonderful friend named Pet and lived in a cool cube house with a strong pyramid roof. Pet even had her own room and a party place for friends!

RHOMBI'S GALLERY

Pet and Rhombi loved 2 dimensional and 3 dimensional shapes of all kinds. Their house was filled with interesting squares, triangles, hexagons, cubes, pyramids and prisms. Rhombi's favorite was a picture of Pet in a trapezoid picture frame. Rhombi loved shapes so much that she was opening a geometric art gallery in town.

RHOMBI'S GALLERY

Rhombi and her friends were taking a leaf walk on their way to visit the site for Rhombi's new business. They found at least five kinds of leaves on the path and printed each one in Rhombi's sketchbook. Suddenly, Rhombi stopped and pointed to the empty and dark contractor's office.

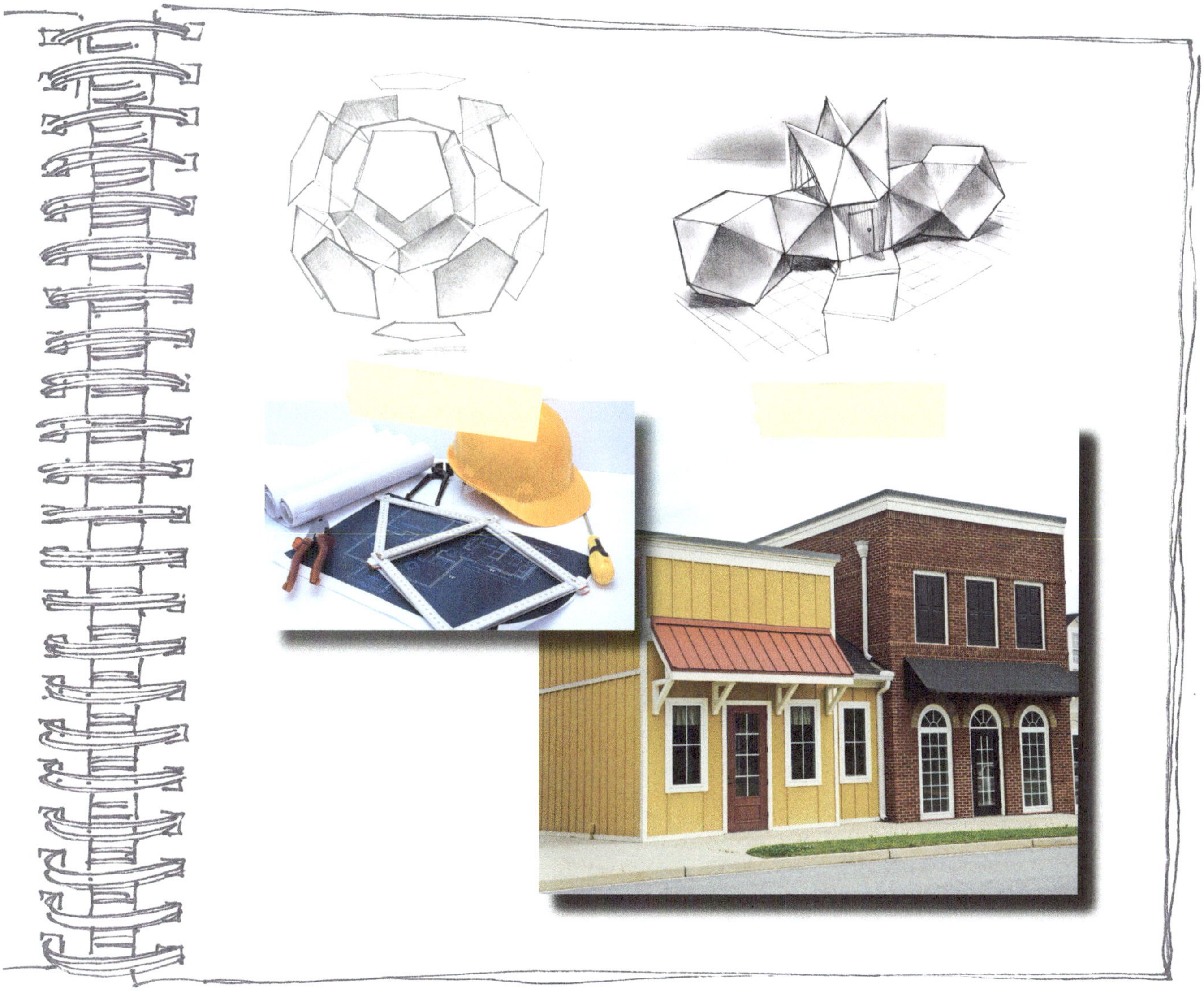

RHOMBI'S GALLERY

Where were her general contractor, engineer, architect, and builder friends? They were all planning to meet for a brainstorm. Rhombi saw toolboxes on a table next to a set of blank blueprints. A note addressed to her was taped to the closed glass door.

RHOMBI'S GALLERY

"Rhombi, I've been called away to fix a building emergency. I can recommend a new architect and designer, or you can wait for me to return. I will be back in 3 weeks."

Rhombi looked at Pet and shook her head.

RHOMBI'S GALLERY

"The gallery fundraiser is set for 2 weeks from now. I need to have scale models for people to see, or I won't be able to raise enough money to build the gallery," said Rhombi. Pet swished her tail and meowed loudly.

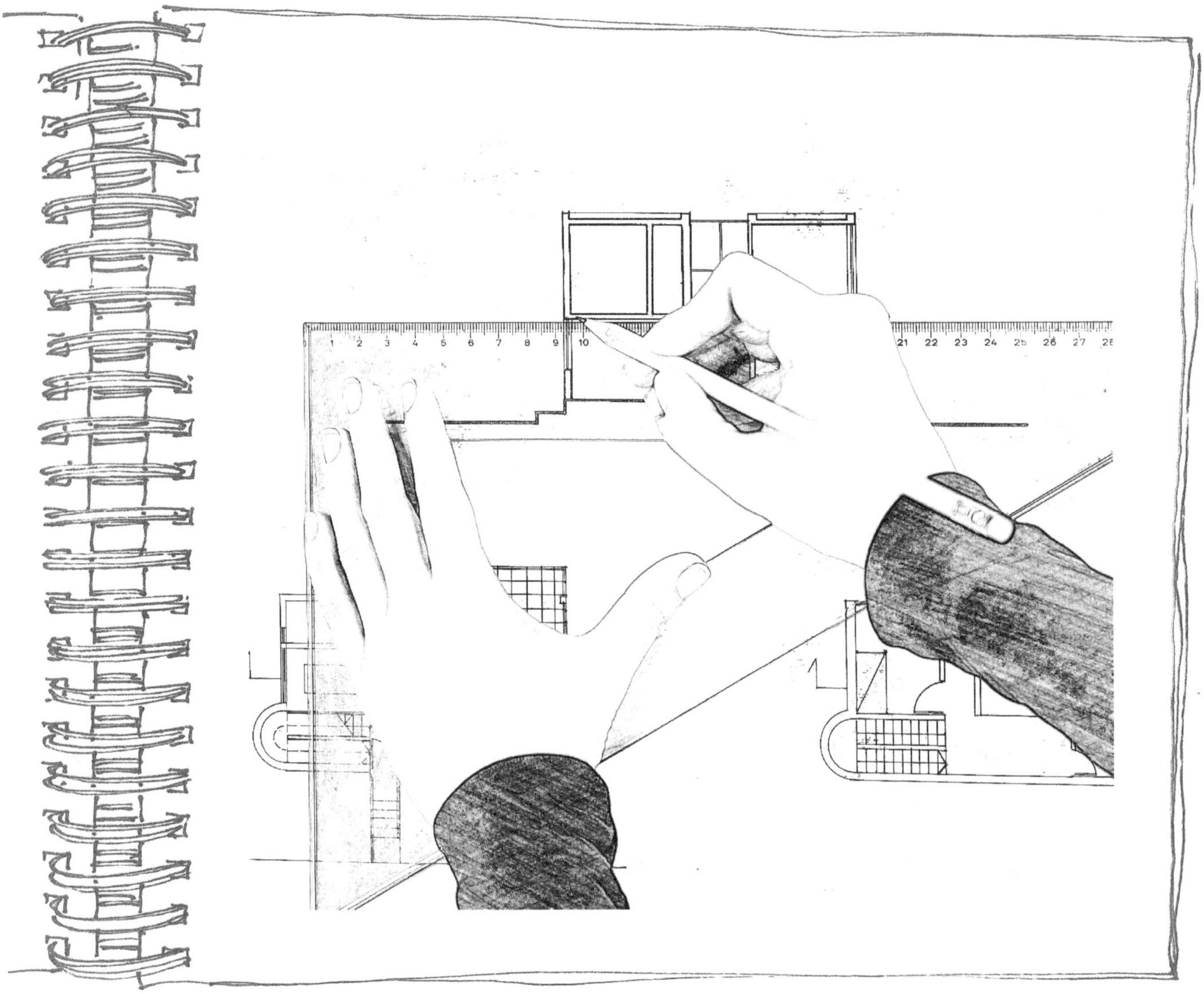

RHOMBI'S GALLERY

"You're right, Pet. I should ask my designer friends for help.""

Rhombi looked out from the pages of her world and asked, "Designers, will you help me create the coolest art gallery you can imagine?"

RHOMBI'S GALLERY
UNIT EXTENSION: TALK ABOUT S.T.E.A.M.

What businesses does a town need to support townspeople?
ex: grocery story or farmer's market, town hall, police department, school, businesses

What would you add to a town to make it really special? ex: ice cream shop, etc.

Make a list of 2D shapes with 3,4,5,6,7,8,9,and 10 sides.
triangle, square (quadrilateral), pentagon, hexagon, heptagon, octagon, nonagon, decagon

Make a list of 3D shapes.
cube, pyramid, prism, cylinder, sphere, etc.

How many different types of leaves (trees) can you name?
ex: oak, hickory, maple, pear, cypress, walnut, cedar, pecan, cherry, aspen, birch, willow

How many cubes could be made from 70 squares? On a number line, move backward in sets of 6 from 70 to zero (do not pass zero). How many whole groups of six did you count? 11

What are blueprints?
Blueprints are designs or technical drawings.

To whom is the letter addressed? Who wrote the letter?
Rhombi and Pet / Quad Rilateral

What is a job for a team?
Designing and building a business to go in the town square requires a team.

What buildings will you choose, and how will you divide the work? Answers will vary.

RHOMBI'S GALLERY
UNIT EXTENSION S.T.E.A.M. CHALLENGE

"Designers, will you help me create an art gallery?"

What type of art will Rhombi sell?
ex: sculptures, framed paintings, murals, mobiles

What kind of building will work best?
ex: Big murals need big space.
Sculptures need special stands and lots of floor space.
Small frame paintings might just need 4 walls.
Mobiles need extra high ceilings.

Draw a design for your building. Make sure the building works for your business.

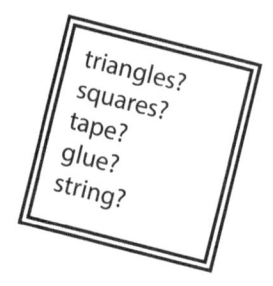

triangles?
squares?
tape?
glue?
string?

On the back of this paper, write out the steps you will need to take.

RHOMBI'S GALLERY
UNIT EXTENSION S.T.E.A.M. RUBRIC

TASK: Students will use their knowledge of habitat, house, pyramids and strength to build a scale model of an art gallery.

	Discussion	Organization	Design & Build
1	• Is well thought out and supports the solution to the challenge or question • Reflects application of critical thinking • Has clear goal that is related to the topic • Is accurate in defining shapes…	• Information is clearly focused in an organized and thoughtful manner • Information is constructed in a logical pattern to support the solution	• Neatly joins and names vertices and edges of 2D shapes to make 3D shapes • Develops a 3D building that reflects thoughtful construction • Building repels water • Building stands up to "dryer" • Building supports weight
2	• Supports the solution • Has application of critical thinking that is apparent • Has no clear goal • Has some factual errors or inconsistencies (i.e. wheels do not rotate around axle)	• Project has a focus but might stray from it at times (more concerned with form than function) • Information appears to have a pattern, but the pattern is not consistently carried out in the project	• Identifies a square but does not combine them to make a cube • Edge seams are mis-joined or uneven • Building does not repel water • Building stands up to "dryer" • Building supports weight
3	• Provides inconsistent information for solution • Has no apparent application of critical thinking • Has no clear goal • Has significant factual errors, misconceptions, or misinterpretations	• Content is unfocused and haphazard • Information does not support the solution to the challenge or question • Information has no apparent pattern	• Does not draw/make 2D shapes • Does not know to join squares to make a cube • Edge seams are missing or ragged • Student project falls apart and student is unable to make changes

APPENDIX

POLYGONS

POLYGON NET PATTERNS 127 - 128
POLYGON TRACING SHAPES 129 - 137
POLYGON TEACHER BACKGROUND 138

STEMVESTIGATIONS ACTIVITY PAGES

RHOMBI'S HOUSE 140 - 144
WEATHER, DESIGNS, STRUCTURES, CUBISM, PERSPECTIVE

UP ON THE ROOF 145 - 149
WATER AND SUN, PYRAMID, ROOF HEIGHT, MATERIALS, EROSION

NEW HOME FOR PET 150 - 154
HABITATS, ATTRIBUTES, FOUNDATIONS, PATTERNS, SIZE AND SCALE

PET'S CELEBRATION 155 - 159
DESSERT, DIGITAL SCALES, STRENGTH, MOSAICS, MEASUREMENT

RHOMBI'S GALLERY 160 - 164
WATERPROOF, STOPWATCH, HEIGHT, GRAPHICS, TIMELINES

MINDBUGS ACTIVITY PAGES

TIME, STOPWATCH, ELAPSED TIME, INCHES, RULERS, WEIGHT,

CORRELATIONS AND ASSESSMENTS

NGSS 176
PRE AND POST ASSESSMENT 177 - 185
WORKS CITED 186
PLANNING PAGES 187 - 188
COMMON CORE CHECKLISTS 189 - 196

APPENDIX: CUBE NET
Print this net for use in a center or during STEMVESTIGATIONS.

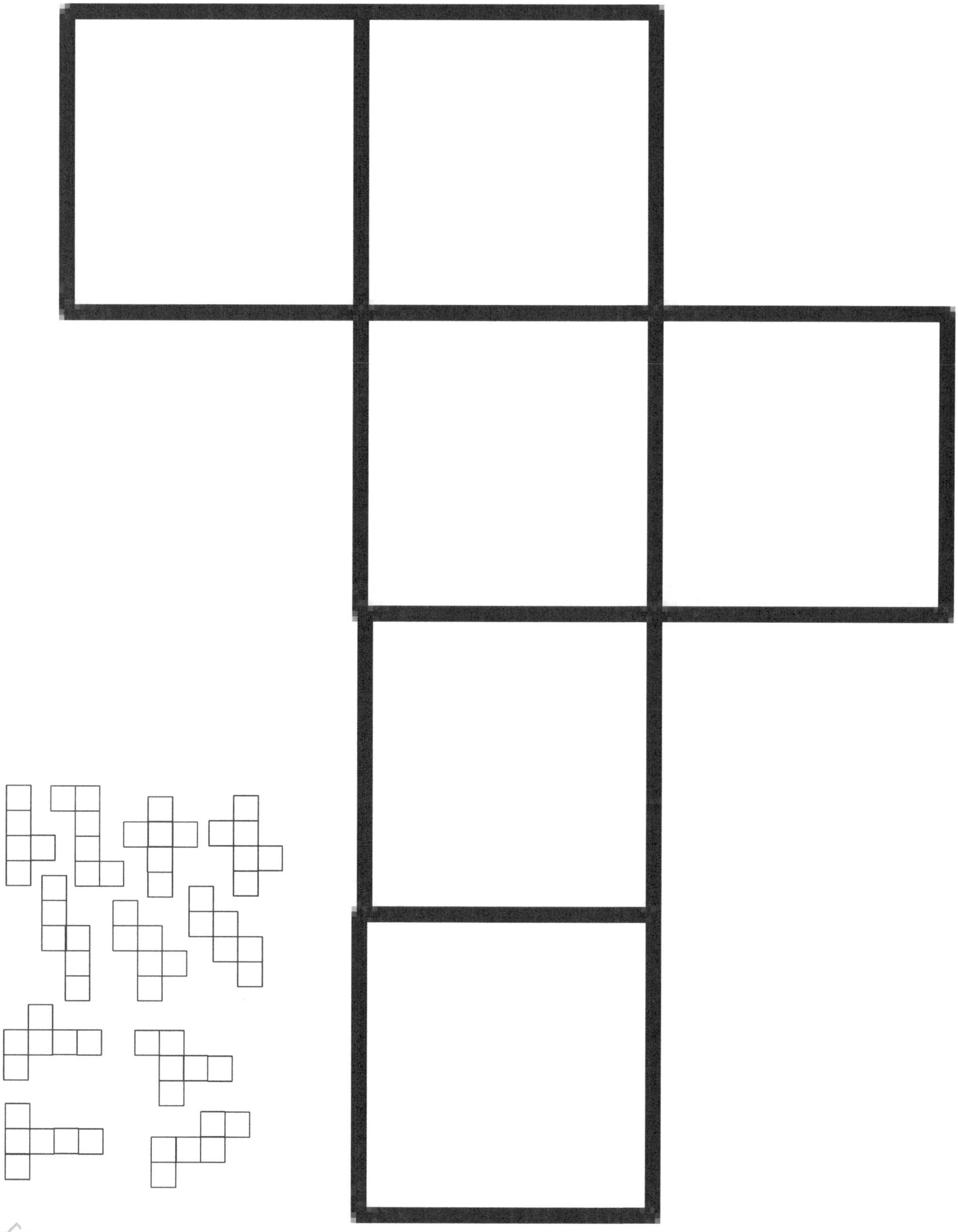

RHOMBOSTEAM 3D SHAPES 2nd Grade   ALL RIGHTS RESERVED ©2014 1O80 EDUCATION, INC.

APPENDIX: PYRAMID NET
Print this net for use in a center or during STEMVESTIGATIONS.

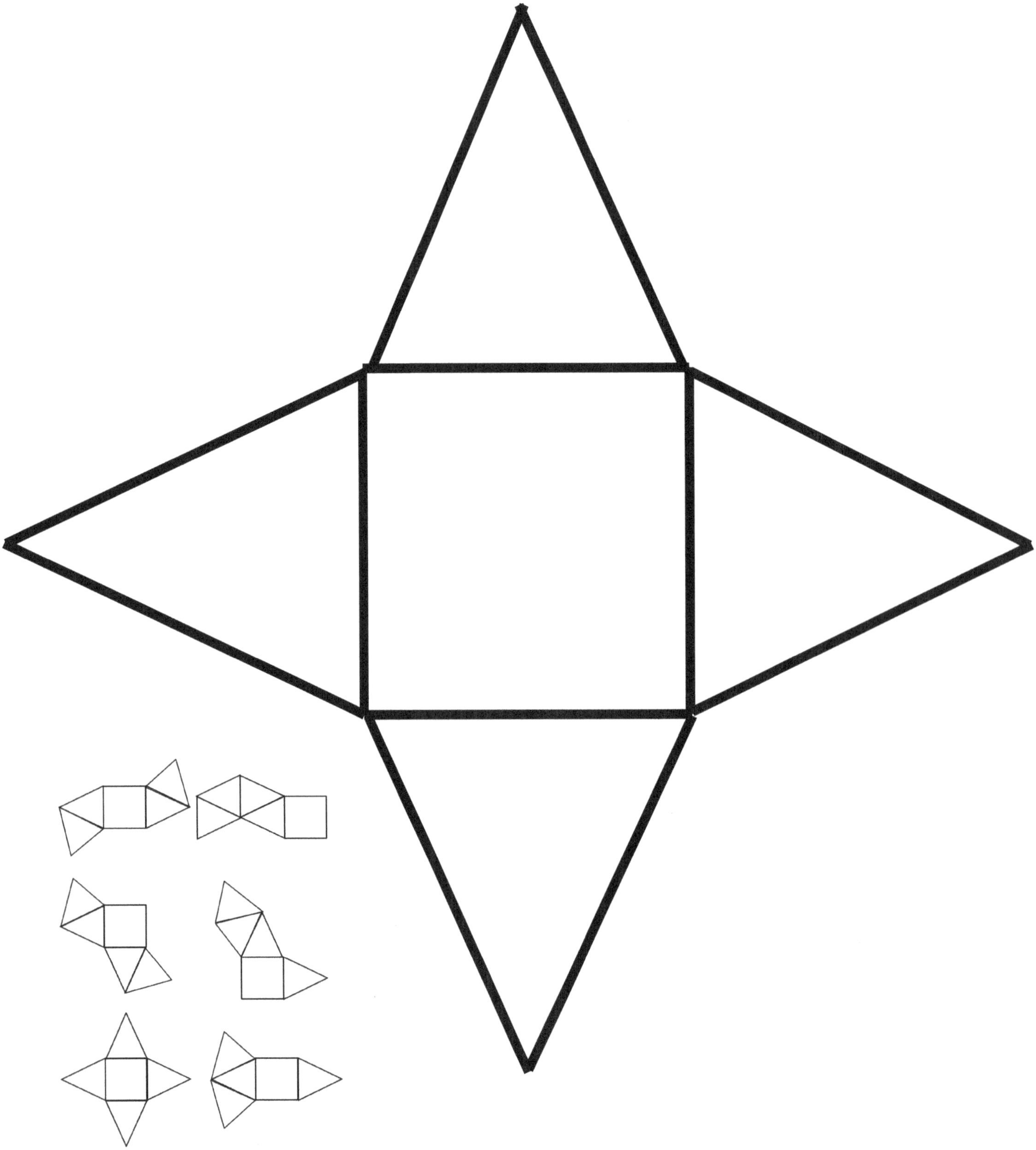

RHOMBI'S POLYGON ROUND-UP
CIRCLE

ABOUT THE CIRCLE
- In the math world, circle refers to the boundary of the shape. "Disk" is used to refer to the whole shape, including the inside.
- A straight line from the center of a circle to the edge is called the radius.
- A straight line that passes from one side of a circle to the other through the center is called the diameter.
- The distance around the outside of a circle is called the circumference. All points on the edge of a circle are the same distance to the center.
- The value of Pi (π) to 2 decimal places is 3.14, it comes in handy when working out the circumference and area of a circle.
- Circles have a high level of symmetry.

RHOMBI'S POLYGON ROUND-UP
TRIANGLE

ABOUT THE TRIANGLE
- Triangles are polygons with the least possible number of sides (three).
- The three internal angles of a triangle always add to 180 degrees.
- An equilateral triangle has three sides of equal length and three equal angles.
- The longest side of a right angle triangle is called the hypotenuse, it is always found opposite the right angle.
- Trigonometry is the study of the relationship between the angles of triangles and their sides.
- Triangle shapes are often used in construction because of their great strength.

RHOMBI'S POLYGON ROUND-UP
RECTANGLE

ABOUT THE SQUARE
- A square is a polygon with 4 sides of equal length and 4 right angle corners (90 degree corners).
- Because it has 4 sides of equal length, a square is a regular quadrilateral.
- A square is also a rectangle with equal sides and a rhombus with right angles.
- The perimeter of a square is 4 times the length of one side.
- Opposite sides of a square are parallel.
- The internal angles of a square add to 360 degrees.
- A square has 4 lines of reflectional symmetry.

RHOMBI'S POLYGON ROUND-UP
PENTAGON

ABOUT THE PENTAGON
- A pentagon is a 5 sided polygon with interior angles that add to 540 degrees.
- Regular pentagons have sides of equal length and interior angles of 108 degrees.
- The US Department of Defense headquarters is named 'the Pentagon'.
- The edible plant okra is shaped like a pentagon.

RHOMBI'S POLYGON ROUND-UP
HEXAGON

ABOUT THE HEXAGON
- A hexagon is a 6 sided polygon with interior angles that add to 720 degrees.
- Regular hexagons have sides of equal length and interior angles of 120 degrees.
- Beehive cells are hexagonal.

RHOMBI'S POLYGON ROUND-UP
HEPTAGON

ABOUT THE HEPTAGON
- A heptagon is a 7 sided polygon with interior angles that add to 900 degrees.
- Regular heptagons have sides of equal length and interior angles of 128.57 degrees.
- The British 50 and 20 pence coins are curved heptagons.

RHOMBI'S POLYGON ROUND-UP
OCTAGON

ABOUT THE OCTAGON
- An octagon is an 8 sided polygon with interior angles that add to 1080 degrees.
- Regular octagons have sides of equal length and interior angles of 135 degrees.

RHOMBI'S POLYGON ROUND-UP
NONAGON

ABOUT THE NONAGON
- A nonagon is a 9 sided polygon with interior angles that add to 1260 degrees.
- Regular nonagons have sides of equal length and interior angles of 140 degrees.

RHOMBI'S POLYGON ROUND-UP
DECAGON

ABOUT THE DECAGON
- A decagon is a 10 sided polygon with interior angles that add to 1440 degrees.
- Regular decagons have sides of equal length and interior angles of 144 degrees.

RHOMBI'S POLYGON ROUND-UP
TEACHER'S BACKGROUND

SUM of INTERNAL ANGLES

GENERAL INFORMATION ON POLYGONS

A polygon is a geometric figure in two dimensions with three or more sides. The name comes from two Greek words, poly, meaning "many," and gon, meaning "angle." A polygon always has as many angles as it has sides. Polygons are named to indicate the number of sides or angles they contain. Thus, a hexagon has six (hexa- means "six") sides and six angles. The sum of a polygon's exterior angles is 360°...
To find the internal angles of a polygon use this formula:
Multiply 180 by (the number of sides minus 2). A hexagon has 6 sides. 6 - 2 = 4 Hexagon is 180° x 4 = 720°

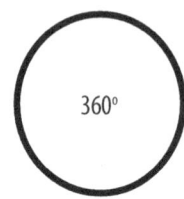
360°

POLYGON TERMINOLOGY
Parts and properties of polygons.

Side: Any one of the straight lines that make up the polygon.

Vertex: A point where any two of the sides of a polygon meet to form an angle.

180°

Angle: A figure formed by the intersection of two sides.

Diagonal: A line that joins any two nonadjacent (not next to each other) vertices.

360°

Perimeter: The sum of the length of all sides.

Area: The space enclosed within the polygon.

540°

TYPES OF POLYGONS

Equilateral: A polygon in which all sides are equal in length.

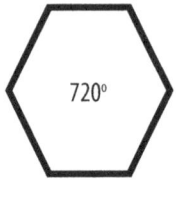
720°

Equiangular: A polygon in which all angles are the same size.

Regular: A polygon that is both equilateral and equiangular. This is the one most often used in schools.

900°

EXAMPLES OF POLYGONS
The most common kinds of polygons include:

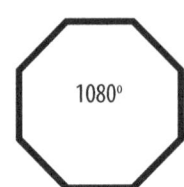
1080°

Parallelogram: A quadrilateral (four-sided figure) in which both pairs of sides are parallel and equal.

Rhombus: A parallelogram in which all four sides are equal.

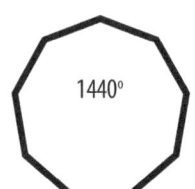
1440°

Rectangle: A parallelogram in which all angles are right angles.

Square: A rectangle in which all four sides are equal.

RHOMBI'S ADVENTURES IN 3D
STEMVESTIGATIONS
MODULE 2: SECOND GRADE

RHOMBI'S HOUSE
STEMVESTIGATION: WEATHER

ACTIVITY ON PAGE 23

Wind measures for the week of _____.

MEASURE THE WIND

Monday-Friday DAY OF THE WEEK	Fahrenheit TEMPERATURE	Circle TODAY'S WEATHER	WIND DIRECTION	WIND SPEED

RHOMBI'S HOUSE
STEMVESTIGATION: DESIGN

ACTIVITY ON PAGE 24

RHOMBI'S HOUSE
STEMVESTIGATION: STRUCTURES

ACTIVITY ON PAGE 25

Make your own tiny house as a scale model. Cut out this shape net to fold a house.

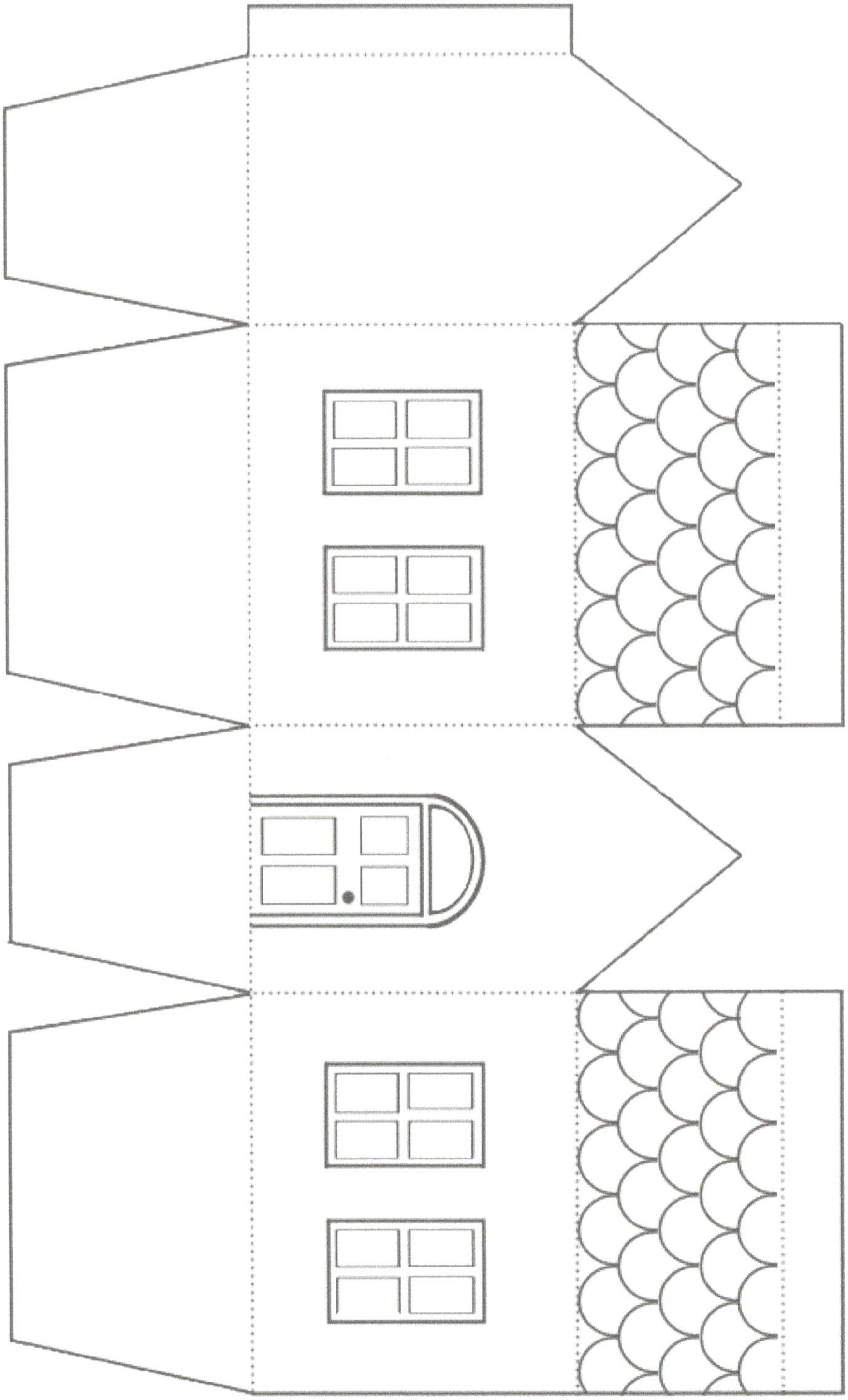

RHOMBI'S HOUSE
STEMVESTIGATION: CUBISM

ACTIVITY ON PAGE 26

Make art with polygons. Use squares and triangles.

Pick a picture with lots of big shapes and colors.
Cut polygons to cover the shapes on your picture.
When you are done, does the art match the picture you pasted?

Paste your picture here.

RHOMBI'S HOUSE
STEMVESTIGATION: PERSPECTIVE

ACTIVITY ON PAGE 27

Draw a stack of cubes.

Build a cube of cubes.
Draw what you see.
Compare your picture with the real cube.

Build your cube of cubes.

Make 3 rows of 3 cubes.
Add a middle layer of 9 cubes.
Add a top layer of 9 cubes.

How many cubes make up your big cube? _____

Draw what you see.

How many cubes are not shown in your drawing?

_____

UP ON THE ROOF
STEMVESTIGATION: WATER EROSION

ACTIVITY ON PAGE 28

Test the effects of water and sun on materials.

Day	Paper	Plastic	Cardboard	Metal
Date:				
Date:				
Date:				
Date:				
Date:				

UP ON THE ROOF
STEMVESTIGATION: PYRAMIDS

ACTIVITY ON PAGE 47

Greetings designers!

Your task today is to design and build a vehicle to move heavy materials from one place to another. Archeologists are building a scale model of the famous Egyptians pyramids. You are hired as a lead engineer in the project. Your vehicle will move the heavy stones from the quarry to the university.

The archeologists will arrive for a meeting in 20 minutes. You must be ready to present your design at that time. Remember to use drawings, charts, and explanations that will help your audience understand your ideas.

Please follow the instructions below and on the next pages of this worksheet. Good Luck!

Sincerely,

Jeannie Ruiz
Jeannie Ruiz, President
Ten80 Elementary, Inc.

INSTRUCTIONS
- You will create a scale model (smaller version) to explain your design.
- The vehicle will move weight for a distance of 3 feet on the table.
- The vehicle should carry at least 6 ounces of sugar cube (pretend stones).

1. <u>Will</u> you move lots of smaller stone blocks, or will you move a few very large blocks?

2. <u>Which</u> is more important in your project: speed or distance?

3. <u>What</u> materials will you use in your model and <u>why</u> do you choose them?

4. <u>How</u> does the wheel and axle or the lever help your project?

UP ON THE ROOF
STEMVESTIGATION: ROOF HEIGHTS

ACTIVITY ON PAGE 48

How tall can you make a tower before it falls?

Use blocks to build a tall, tall tower.

1. Choose the blocks to use, or make your own cubes.
2. Stack and count the blocks until the tower falls.
3. Record the number of blocks. Repeat 4 times.

TEST	# BLOCKS
1	9 blocks
2	
3	
4	

Draw your tower.

Try building a pyramid as tall as your tower.
1. How many layers of blocks did you use? _____
2. How many blocks are in your pyramid? _____
3. How many more blocks did your pyramid need? _____
4. Which building is more stable?

Draw your pyramid.

UP ON THE ROOF
STEMVESTIGATION: MATERIALS

ACTIVITY ON PAGE 50

Make a clay brick for building.

Find the best mix of materials.

Bricks may be made when clay soil is mixed with straw. Mix 3 cups of clay soil with water until it becomes quite thick. Chop straw into tiny bits (or have an adult use a food processor). Use your hands to squeeze the clay and straw together.

QUESTIONS TO INVESTIGATE
- How much straw should be used?
- How much water?
- What is the best mixture to make a clay shape that dries without crumbling?
- How will you dry the clay brick?

TEST #	AMOUNT OF SOIL	AMOUNT OF WATER	AMOUNT OF STRAW	SECONDS TO MIX
1	CUPS	TABLESPOONS	TABLESPOONS	SECONDS
2				
3				
4				
5				

Make a scale model of a pyramid once you've found the best mixture for a brick. Leave the pyramid in full sun for 2 days to bake the clay.

UP ON THE ROOF
STEMVESTIGATION: EROSION

ACTIVITY ON PAGE 51

Design a drip system to test erosion.

Name some things that drip.

1.
2.
3.

What makes these things drip?

1.
2.
3.

How could you make a drip tool using what you already know?

What materials will you need to build your drip tool?

Remember, the "dripper" needs to drip drops of water for at least 1 hour before refilling. The water must drip very slowly over a long time.

RHOMBOSTEAM 3D SHAPES 2nd Grade   ALL RIGHTS RESERVED ©2014 1O80 EDUCATION, INC.

NEW HOME FOR PET
STEMVESTIGATION: HABITATS

ACTIVITY ON PAGE 69

Make an animal show-and-tell.

Create a home for your favorite toy animal.
1. Bring a stuffed or plastic animal to class.
2. Learn about the real version of your animal

Kind of Animal	
Habitat	
Food	
2 or 3 Facts	
Size in Inches	

Which polygon shape makes the best home for your animal? Design a habitat for your toy.

List the materials you need.

Present the animal, its geometric habitat, food source, and facts during a show-and-tell.

NEW HOME FOR PET
STEMVESTIGATION: ATTRIBUTES

ACTIVITY ON PAGE 70

Make a list of things you love, like, and want to do.
Then, design and draw a "wild new you."

Things I love to eat... *example: apples* 	How do I eat them? *example: teeth*
Things I want to do... *example: fly* 	What would I need? *example: wings* 
Favorite activities... *example: cooking*	What would I need? *example: hands*

On the back of this page, draw a "wild you" that covers the items on your list of likes and wants.

NEW HOME FOR PET
STEMVESTIGATION: STRUCTURES & FOUNDATIONS

ACTIVITY ON PAGE 71

Make a floor that holds more weight than a textbook!

Offer educated guesses.
Make scale models.
Test your hypothesis.
Look at the data.
Repeat!

Collect your materials.
CUPS   CARDBOARD   BOOKS

Test 1 Cup 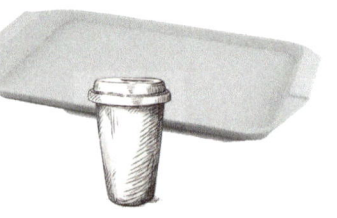   How many books? _____

Test 2 Cups 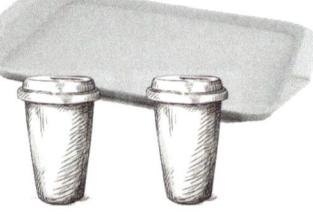   How many books? _____

Test 3 Cups 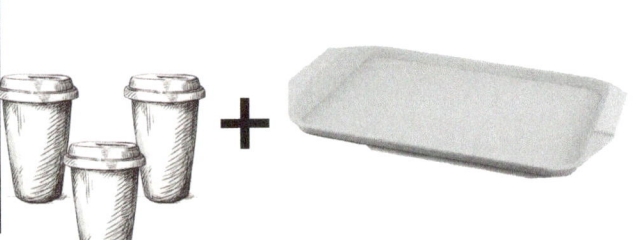   How many books? _____

NEW HOME FOR PET
STEMVESTIGATION: PATTERNS

ACTIVITY ON PAGE 72

Design a tesselation. Your pattern should repeat all over the page. What shape will you use?

NEW HOME FOR PET
STEMVESTIGATION: SIZE AND SCALE

ACTIVITY ON PAGE 73

Create a scale to help you talk about size. Will you use animals, plants, or objects as your guide?

SIZE	FROM	TO
TINY		
LITTLE		
SMALL		
AVERAGE		
BIG		
LARGE		
GIANT		

PET'S CELEBRATION
STEMVESTIGATION: DESSERT

ACTIVITY ON PAGE 91

Create a tool for making mashed fruit desserts.

What shape is best for cutting? What shape is best for mashing? Your teeth are a clue. Teeth in the back of your mouth crush and mash. Teeth in the front of your mouth cut and tear. Which shape will you use to make a creamy frozen snack?

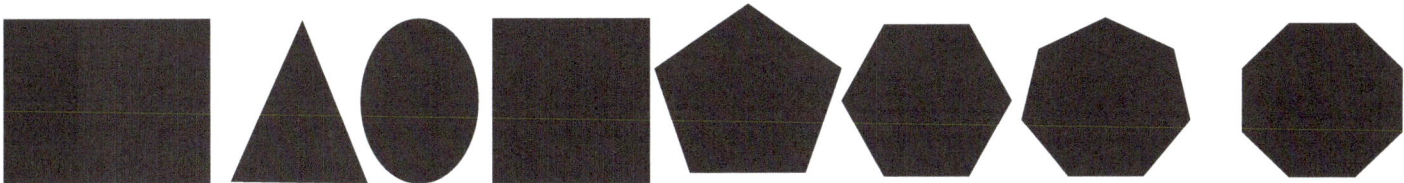

Have you seen a potato masher? Is it solid, or does it have holes?

Draw your tool design.

Draw or list materials you will need.

PET'S CELEBRATION
STEMVESTIGATION: DIGITAL SCALES

ACTIVITY ON PAGE 92

Use a digital scale to weigh 6 objects.

OBJECT	MY ESTIMATE	WEIGHT
CRAYON		ounces
MUG		ounces
SHOE		ounces
BOOK		ounces
?_____		ounces

5 grams
and no parts of another gram

PET'S CELEBRATION
STEMVESTIGATION: STRENGTH

ACTIVITY ON PAGE 93

Build a cylinder that supports a 1-pound object.

| Draw a cylinder. | Make a cylinder from one piece of paper and tape. What material will you choose?
• lined paper
• construction paper
• cardstock
• newspaper
• OTHER? |

What will you do to make your cylinder strong enough to support a 1-pound object?

FILL 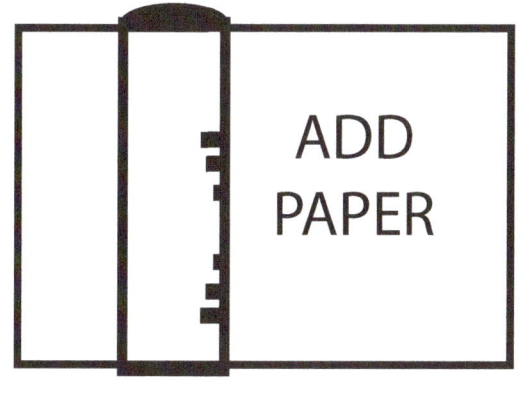 ADD PAPER  CHANGE SIZE

RHOMBOSTEAM 3D SHAPES 2nd Grade   ALL RIGHTS RESERVED ©2014 1O80 EDUCATION, INC.

PET'S CELBRATION
STEMVESTIGATION: MOSAICS

ACTIVITY ON PAGE 94

Make art with polygons. Use squares and triangles.

Draw a picture. Use white crayon and black paper. Cut the paper into strips to make long rectangles. Cut rectangles into squares. Cut squares in half from point to point to make triangles. Fill your picture with these polygons to make a mosaic.

PET'S CELEBRATION
STEMVESTIGATION: MEASUREMENT / LENGTH

ACTIVITY ON PAGE 95

Measure string to make line art.

Lay string along the length of a measuring stick. Cut 36 inches (3 feet) of string. Use your string to make a cool shape. Glue the shape to paper.

RHOMBI'S GALLERY
STEMVESTIGATION: WATERPROOF

ACTIVITY ON PAGE 109

Test the effects of water and sun on materials.

Day	Paper	Plastic	Cardboard	Metal
Date:				
Date:				
Date:				
Date:				
Date:				

RHOMBI'S GALLERY
STEMVESTIGATION: STOPWATCH

ACTIVITY ON PAGE 110

Use a stopwatch to time yourself.

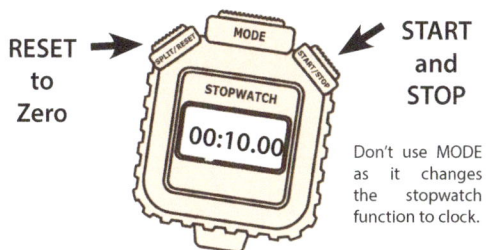

RESET to Zero

START and STOP

Don't use MODE as it changes the stopwatch function to clock.

Say animals for 10 seconds. Try 3 times.
1. How many animals did you say?____
2. How many animals did you say?____
3. How many animals did you say?____

Say fruits for 10 seconds. Try 3 times.
1. How many fruits did you say?____
2. How many fruits did you say?____
3. How many fruits did you say?____

Say the alphabet. How many seconds pass?
1. How many seconds?____
2. How many seconds?____
3. How many seconds?____

Enter your data in this chart. Talk about the chart with your team.

	Animals	Fruits
1		
2		
3		

RHOMBI'S GALLERY
STEMVESTIGATION: TALLEST TOWER

ACTIVITY ON PAGE 111

Use newspapers to make a tall, tall tower.

Stack. Fold. Roll. Turn the newspaper into building material.

  Design a tower.

Draw a tower design.   Draw another tower design.

<------
Pick
One
------>

Measure height. How tall is your tower?   _____ inches

How many sheets of papers did you use?   _____ inches

If you could redesign your tower, what would you change? Draw your new tower. Tell a friend why you want to make changes.

RHOMBI'S GALLERY
STEMVESTIGATION: GRAPHICS

ACTIVITY ON PAGE 112

Create a logo. Make a picture that represents you!

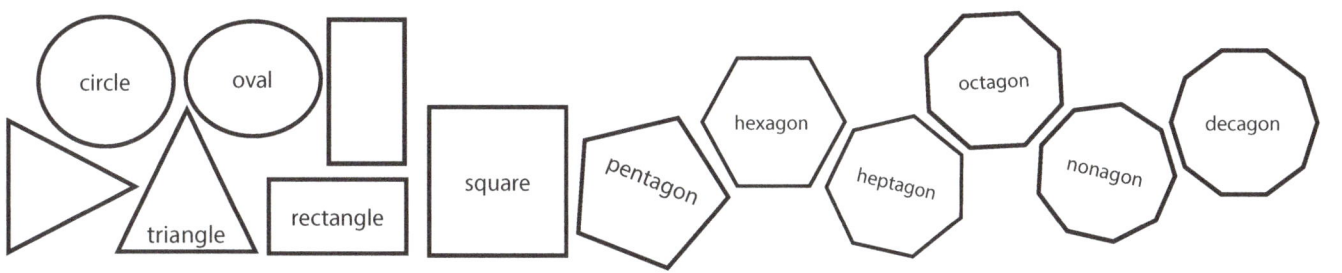

red   yellow   blue   pink   black   gray
orange   green   purple   white   brown

1. Pick a polygon.   2. Pick a color.   3. Pick a thing that fits you.

example

Create your logo.

RHOMBI'S GALLERY
STEMVESTIGATION: TIMELINES

ACTIVITY ON PAGE 112

Look at your week. Look at a friend's week. Compare.

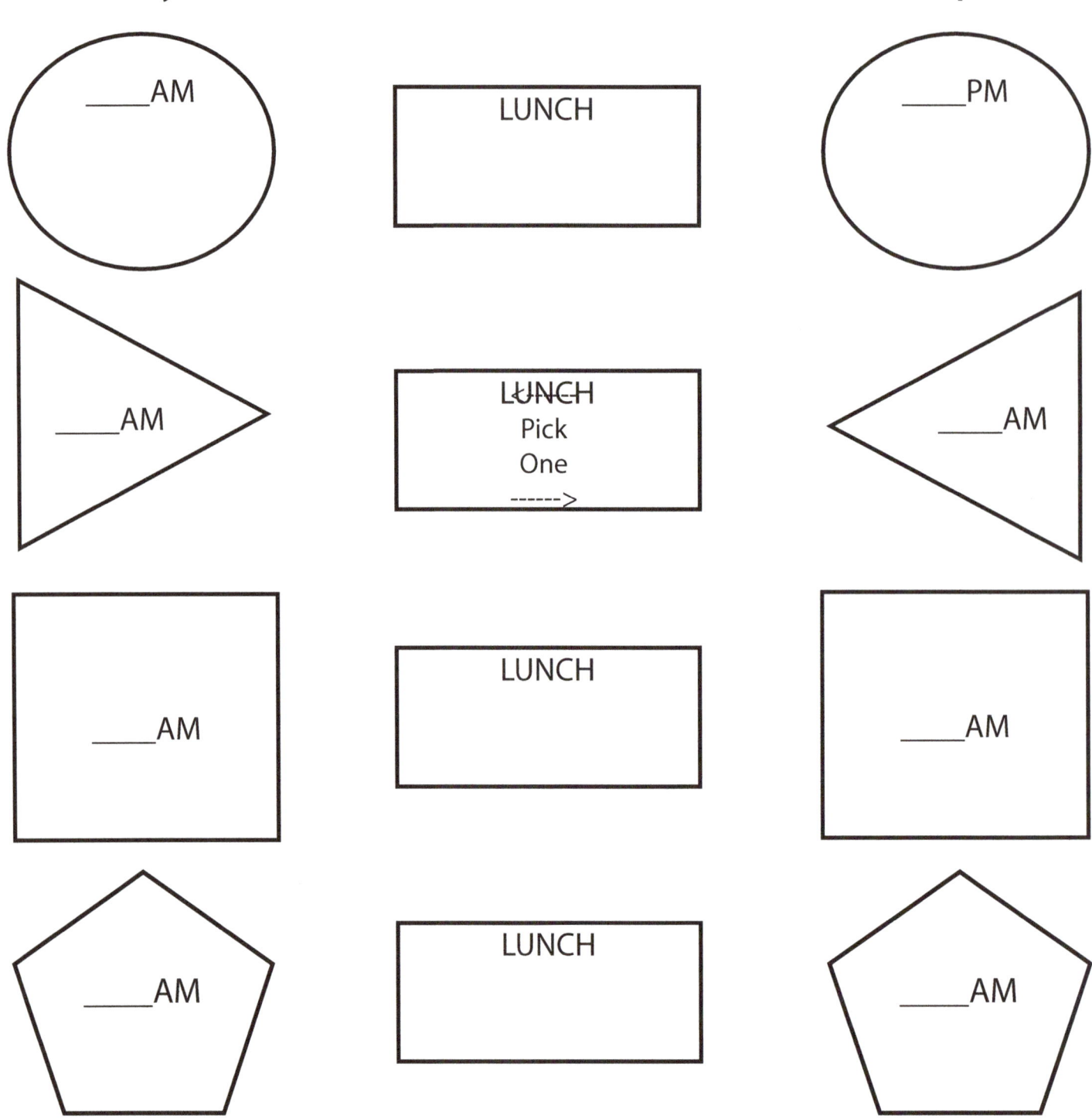

TIME
    TIME   167
    STOPWATCH   168
    ELAPSED TIME   169
    REACTION TIME   170

LENGTH
    RULER AND TAPE MEASURE   171
    INCHES AND YARDS   172
    PRACTICE MEASURING LENGTH   173

WEIGHT
    USE A DIGITAL SCALE   174
    PRACTICE USING A SCALE   175

RHOMBI'S ADVENTURES IN 3D
MINDBUGS IN MEASUREMENT

MODULE 2: SECOND GRADE

I understand elapsed time.

• •

Instructions: Use your stopwatch to find your personal measures. How many of each activity can you complete in 30 seconds?

This stopwatch shows 3 seconds and 25 hundredths of another second.

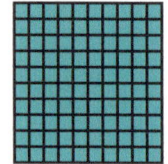 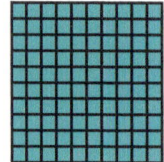 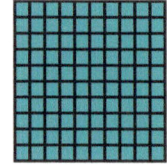 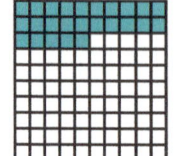

Activity	Time in Seconds	Time in Seconds
Jumping Jacks		
Sit Ups		
Snap Your Fingers		
Clap Your Hands		

I will understand the stopwatch.

* Start and stop the timer by pressing the same button, Start 1 as shown here.

* Reset the timer to zero by pressing the opposite button, Start 2 as shown here.

Practice Reading Your Stopwatch

The stopwatch shown here displays
hours : minutes : seconds. hundredths of a second

Read your stopwatch using the decimal point. Any number after the last decimal point is part of a second, called a fraction of a second. This stopwatch shows two numbers to the right of the decimal point. These numbers represent hundredths of a second.

This watch shows 2 seconds and 34 hundredths of a second.
This value represents two full seconds and one fraction of a second.

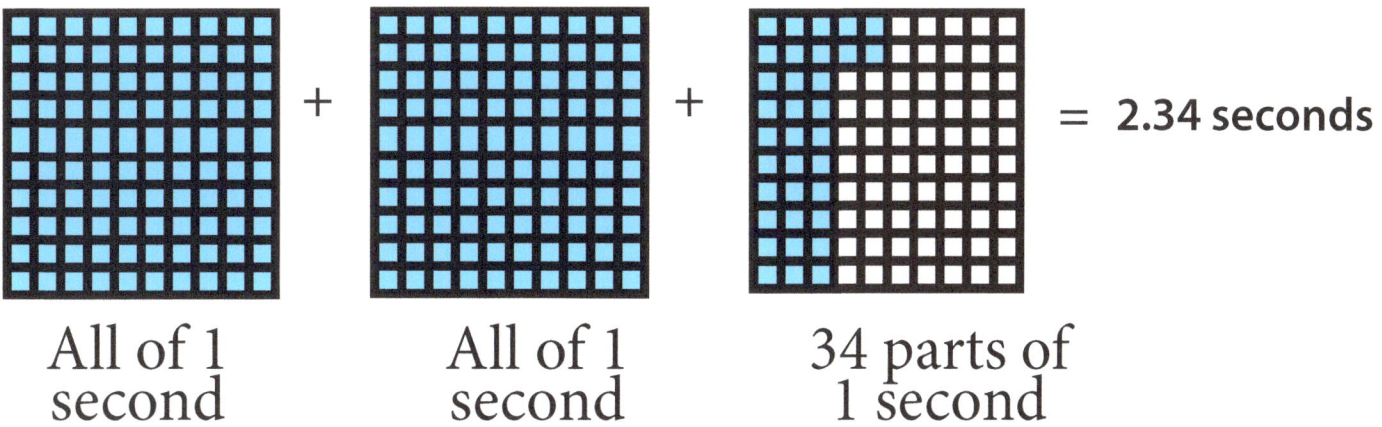

All of 1 second + All of 1 second + 34 parts of 1 second = 2.34 seconds

I will understand elapsed time.

• •

Instructions: Use your stopwatch to find your personal measures. How many of each activity can you complete in 30 seconds?

This stopwatch shows 3 seconds and 25 hundredths of another second.

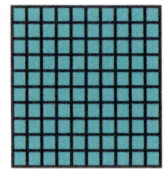 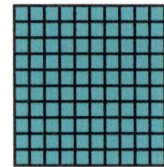 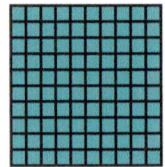 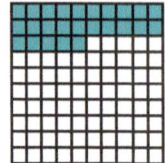

Activity	Time in Seconds	Time in Seconds
Jumping Jacks		
Sit Ups		
Snap Your Fingers		
Clap Your Hands		

I will understand reaction time.

How fast is an eye blink?

From measurements taken over time, we know that an eye blink averages between .10 and .15 seconds. This means that if your reaction time (how fast you can stop and start the stopwatch) is greater than 0.15 seconds, the blink is over before you can push the buttons and take a reading.

Because people blink, the average person spends about 23 minutes in darkness every day. This time in darkness may be why the brain has learned to "fill in the blank spaces" for us on occasion.

Can you use a stopwatch to time an eye blink? Try it..

I will understand the ruler and measuring tape.

LEARN TO READ A TAPE MEASURE.

One side of the tape measure or ruler shows customary units.

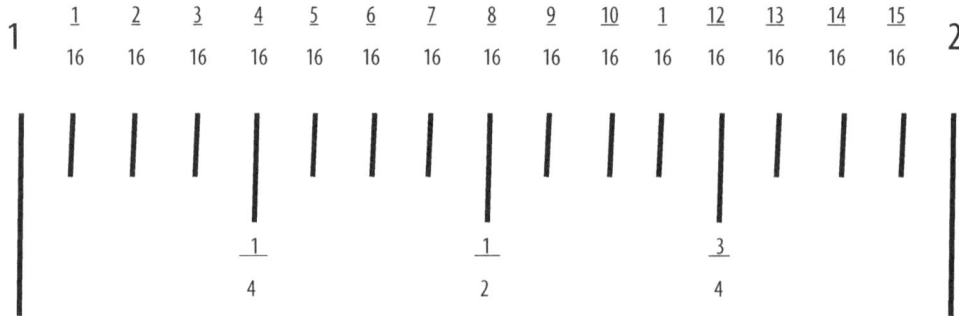

One side of the tape measure or ruler shows metric units.

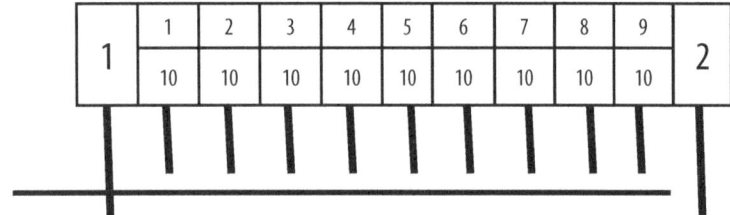

On both sides of the tape measure or ruler, the following is true:

The darker, longer lines show whole numbers like 1,2,3 or 4.

The smaller, shorter lines show parts of a whole number.

I will understand the inch and yard.

• •

INCH BY INCH

Let your fingers do the walking when it comes to measurement. For thousands of years, the human body has acted as a ruler for measuring. A finger's or hand's width and the length of your foot were all accepted ways to measure length.

In the time of England's Round Table, when knights in armor fought battles on horseback, an inch was just the width of a man's thumb. How confusing that must have been! Compare your thumb to someone else's in the room, and imagine trying to build a castle using all the different measurements of a thumb's width.

King Edward of England realized that a constant distance was important to trade. He decreed that the inch should be one thirty-sixth of a yard. A permanent measuring stick made of iron served as a master standard yardstick for the entire kingdom.

Americans decided that the inch and yard would be measured against the European metric system. America created a bronze bar measuring 82 inches. The bar has been replaced at least once since it's first use in 1832.*

*http://www.cftech.com/BrainBank/OTHERREFERENCE/WEIGHTSandMEASURES

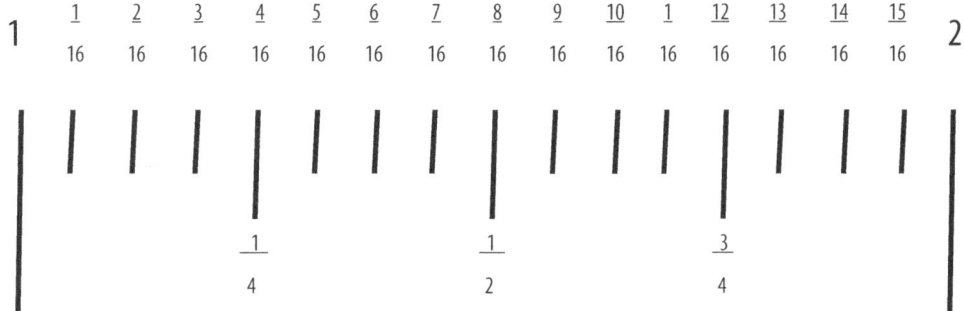

I will understand length.

Measure your right arm from index finger to shoulder.

_____in

Measure your left arm from index finger tip to shoulder.

_____in

What is your arm span? Measure the distance from hand to hand with arms outstretched.

_____in

Did you know that it's perfectly normal for one arm to be a little bit shorter than the other arm?

I will understand tools for measuring weight and mass.

WHAT IS A DIGITAL SCALE?

A scale is a tool used to measure the amount of something. When an object is placed on the scale, the display screen will show you how much the object weighs.

FIND A SPECIFIC WEIGHT

Place the scale on a table. Press the {ON/TARE} switch.
The numbers "0000" or "----" will appear on the display.
Wait for the display to read "0" before the scale is ready to use.
Place an object on the weighing tray.

Decide how much of something you want to weigh.
Do you want to find 1 ounce, 5 ounces, or 10 ounces.
Place an estimated number of weights on the tray, and read the display. You may have to add a bit, or take some out in order to get the exact number you want. When the display reads "100," you have found the correct number of items to meet your specific weight.

FIND THE UNKNOWN WEIGHT OF A GIVEN OBJECT

Wait for the display to read "0" and place the object on the scale.
Press "units" to choose "ounces." Read the display from left to right.

I will understand tools for measuring weight and mass.

DIGITAL SCALES

Find the weight of each writing tool in ounces.

_____ ounces

pencil

_____ ounces

pen

_____ ounces

ruler

_____ ounces

tape

RHOMBI'S ADVENTURES IN 3D
NGSS KINDERGARTEN * FIRST GRADE * SECOND GRADE CHECKLIST

NGSS K 2		Module STEMvestigations Engineering Design Project	GIFTS	MODULE 2 BUILDING RHOMBIS HOME				
				House	Roof	Pet	Party	Innovate
K-2	K-2.Engineering Design							
	K-2-ETS1-1	Ask questions, make observations, and gather information about a situation people want to change to define a simple problem that can be solved through the development of a new or improved object or tool.	X	X	X	X	X	X
	K-2-ETS1-2	Develop a simple sketch, drawing, or physical model to illustrate how the shape of an object helps it functions as needed to solve a given problem.	X	X	X	X	X	X
	K-2-ETS1-3	Analyze data from tests of two objects designed to solve the same problem to compare the strengths and	X	X	X	X	X	X
K	Forces and Interactions: Pushes and Pulls							
	K-PS2-1	Plan and conduct an investigation to compare the effects of different strengths or different directions of pushes and pulls on the motion of an object.		X			X	
	K-PS2-2	Analyze data to determine if a design solution works as intended to change the speed or direction of an object with a push or a pull.*		X			X	
K	Interdependent Relationships in Ecosystems: Animals, Plants, and Their Environment							
	K-LS1-1	Use observations to describe patterns of what plants and animals (including humans) need to survive						
	K-ESS2-2	Construct an argument supported by evidence for how plants and animals (including humans) can change the environment to meet their needs						
	K-ESS3-1	Use a model to represent the relationship between the needs of different plants or animals (including humans)		X	X	X	X	
	K-ESS3-3	Communicate solutions that will reduce the impact of humans on the land, water, air, and/or other living things in the local environment.		X	X	X	X	
K	Weather and Climate							
	K-PS3-1	Make observations to determine the effect of sunlight on Earth's surface.		X			X	
	K-PS3-2	Use tools and materials to design and build a structure that will reduce the warming effect of sunlight on an area.		X	X		X	
	K-ESS2-1	Use and share observations of local weather conditions to describe patterns over time.		X	X	X	X	
1	Waves: Light and Sound							
	1-PS4-1	Plan and conduct investigations to provide evidence that vibrating materials can make sound and that sound can make materials vibrate.						
	1-PS4-2	Make observations to construct an evidence-based account that objects can be seen only when illuminated.						
	1-PS4-3	Plan and conduct an investigation to determine the effect of placing objects made with different materials in the path of a beam of light.						
	1-PS4-4	Use tools and materials to design and build a device that uses light or sound to solve the problem of communicating over a distance.						
1	Structure, Function, and Information Processing							
	1-LS1-1	Use materials to design a solution to a human problem by mimicking how plants and/or animals use their external parts to help them survive, grow, and meet their needs.		X	X	X	X	
	1-LS1-2	Read texts and use media to determine patterns in behavior of parents and offspring that help offspring survive.						
	1-LS3-1	Make observations to construct an evidence-based account that young plants and animals are like, but not exactly like, their parents.						

RHOMBI'S ADVENTURES IN 3D
PRE AND POST ASSESSMENT PAGES

PRINT FOR EACH STUDENT

RECOGNIZE SHAPES	178
REPRODUCE SHAPES	179
MOVE FROM 2D TO 3D	180
UNDERSTAND EQUALITY	181
MOTOR SKILLS	182

PRINT ONE SET FOR TEACHER

NUMERACY, CARDINALITY	183
OBJECT PERMANENCE	184
RUBRIC FOR NUMERACY, CARDINALITY, OBJECT PERMANCE	185

RHOMBI'S ADVENTURES IN 3D
RECOGNIZE SHAPES

Color the square.	□	⬠	△	○
Color the triangle.	□	⬠	△	○
Color the rectangle.	▭	⬠	△	○
Color the circle.	□	⬠	△	○
Color the pentagon.	▭	⬠	△	○

RHOMBI'S ADVENTURES IN 3D
REPRODUCE SHAPES

Draw a square.	
Draw a triangle.	
Draw a rectangle.	
Draw a circle.	
Draw a pentagon.	

RHOMBI'S ADVENTURES IN 3D
MOVE FROM 2D TO 3D

Mark the shape you see on a cube.

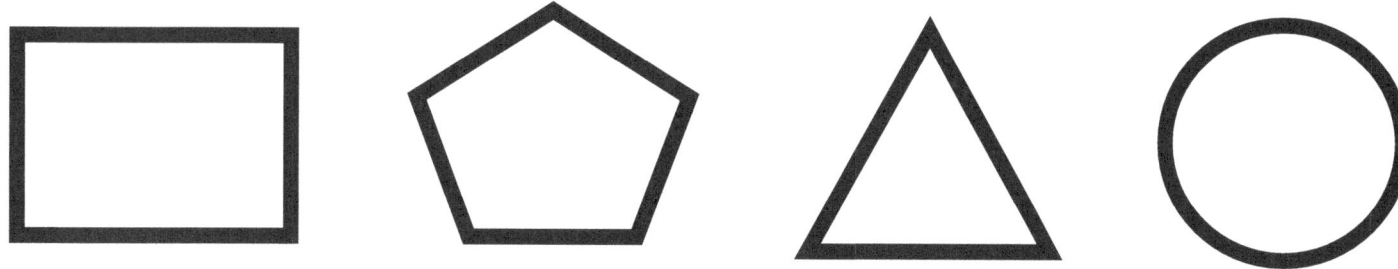

Mark the shape you see on a pyramid.

Mark the shape you see on a the end of a cylinder.

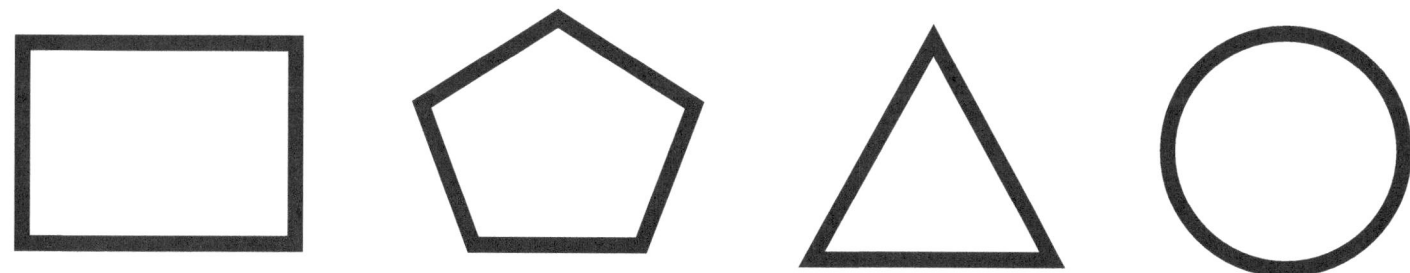

RHOMBI'S ADVENTURES IN 3D
UNDERSTAND EQUALITY

Draw a line to break each shape into two equal parts.

Draw the other half of each shape.

RHOMBI'S ADVENTURES IN 3D
RECOGNIZE SHAPES IN THE WORLD

Glue the cut shapes to this page to make a picture.

RHOMBI'S ADVENTURES IN 3D
ASSESS NUMERACY & CARDINALITY

DO NOT COPY FOR STUDENT
one-on-one with teacher

Show the child 2 pennies. Ask, "How many objects are in my hand?"

Child immediately says, "2." Has solid sense of abstract number 2...
Ready for discussions without visuals...

Child counts out loud and answers "2." Has basic understanding of abstract number 2...
Ready for handouts...

Child points to each object, counting out loud and answers "2."
Needs concrete objects to count while transitioning to abstract number sense...

Repeat actions for numbers 3 and 4.

Recognizing the sum of a pile of objects beyond 4 is almost impossible. Try it yourself. You will reflexively break the larger number into two or more lesser groups before combining to say the greater number.

All students will break numbers beyond 4 into smaller "piles." Note which students count by ones beyond 4 and which students automaticallly do some mental addition to give you the whole number.

RHOMBI'S ADVENTURES IN 3D
ASSESS OBJECT PERMANENCE

DO NOT COPY FOR STUDENT
one-on-one with teacher

Show the child your hand as you fold 2 pennies into your closed fist. Reopen the same hand. Ask. How many objects are in my hand?"

Child immediately says, "2." Has solid sense of object permanence for 2...
Ready for discussions without visuals...

Child counts out loud and answers "2." Has basic understanding of object permance for 2.
Ready for handouts...

Child points to each object, counting out loud and answers "2."
Needs concrete objects to count while transitioning to object permanence...

Repeat actions for numbers 3 - 9.

Remember, in this assessment, your goal is to determine a student's grasp of object permanence NOT ability to count. The outcome is dependent on the student's ability to restate the cardinal number.

RHOMBI'S ADVENTURES IN 3D
ASSESS NUMERACY & CARDINALITY

Print rubric for your records. Perform assessment as early as possible.

STUDENT NAME	1	2	3	4	5	6	7	8	9

NOTE: All students will need to break amounts into smaller groups in order to recognize the cardinal numbers beyond "4." The brain naturally grasps groups of 1,2,3,4. It's why phone numbers are broken into smaller groups.

RHOMBI'S ADVENTURES IN 3D
REFERENCE MATERIALS AND SUGGESTED READING

"Circle." Wikipedia, the Free Encyclopedia. Wikimedia Foundation, Inc, n.d. Web. 7 July 2014.

Dehaene, Stanislas. The Number Sense: How the Mind Creates Mathematics. New York: Oxford UP, 1997. Print.

Devlin, Keith J. The Math Gene: How Mathematical Thinking Evolved and Why Numbers Are Like Gossip.
    New York: Basic Books, 2000. Print.

Ferris Bueller's Day Off. Dir. John Hughes. Perf. Matthew Broderick, Mia Sara, Alan Ruck, Jennifer Grey,
    and Jeffrey Jones. Paramount, 1986. Film.

Feynman, Richard P, and Jeffrey Robbins. The Pleasure of Finding Things Out:
    The Best Short Works of Richard P. Feynman. Cambridge: Perseus Books, 1999. Print.

Furoy, Michael, Rosemary T. Wong, and Harry K. Wong. The Effective Teacher. Mountain View:
    Harry K. Wong Publications, Inc., 2009. Print.

Ifrah, Georges. From One to Zero: A Universal History of Numbers. New York: Viking, 1985. Print.

Lee, Joon S., and Herbert P. Ginsburg. "Preschool Teachers' Beliefs About Appropriate Early Literacy and
    Mathematics Education for Low and Middle-Socioeconomic Status Children." Early Education
    and Development 18.1 (2007):   111-143. Print.

Ma, Liping. Knowing and Teaching Elementary Mathematics: Teachers' Understanding of Fundamental
    Mathematics in China and the United States. Hillsdale, NJ: Lawrence Erlbaum Associates, 1999. Print.

Mathemagic. Chicago: World Book—Childcraft International, Inc., 1980. Print.

McLeish, John. The Story of Numbers. New York: Fawcett Columbine, 1994. Print,

Next Generation Science Standards. N.p., n.d. Web. 7 July 2014.

Wassily Kandinsky – biography, paintings, books. N.p., n.d. Web. 7 July 2014.

"Inquiry Based Science: What Does It Look Like??" Connect Magazine. 1995: 13. Print.

RHOMBI'S ADVENTURES IN 3D
TEACHER'S PRINTABLE UNIT PLANNING PAGE

3D SHAPES: UNIT _____					
STEMVESTIGATION	DATE	TIME	WHAT I'VE GOT	WHAT I NEED	PLANNING NOTES
SCIENCE					
TECHNOLOGY					
ENGINEERING					
ARTS					
MATH					
CHALLENGE					

Common Core Planning

Checklists for STEM Success

SECOND GRADE MATHEMATICS

SECOND GRADE MATH

Operations & Algebraic Thinking

2.NBT.3.	Read and write numbers to 1000 using base-ten numerals, number names, and expanded form.	Understand place value.
2.NBT.4.	Compare two three-digit numbers based on meanings of the hundreds, tens, and ones digits, using >, =, and < symbols to record the results of comparisons.	
2.NBT.5.	Fluently add and subtract within 100 using strategies based on place value, properties of operations, and/or the relationship between addition and subtraction.	Use place value understanding and properties of operations to add and subtract.
2.NBT.6.	Add up to four two-digit numbers using strategies based on place value and properties of operations.	
2.NBT.7.	Add and subtract within 1000, using concrete models or drawings and strategies based on place value, properties of operations, and/or the relationship between addition and subtraction; relate the strategy to a written method. Understand that in adding or subtracting three-digit numbers, one adds or subtracts hundreds and hundreds, tens and tens, ones and ones; and sometimes it is necessary to compose or decompose tens or hundreds.	
2.NBT.8.	Mentally add 10 or 100 to a given number 100–900, and mentally subtract 10 or 100 from a given number 100–900.	
2.NBT.9.	Explain why addition and subtraction strategies work, using place value and the properties of operations.	

SECOND GRADE MATH

Operations & Algebraic Thinking

2.OA.1.	Use addition and subtraction within 100 to solve one- and two-step word problems involving situations of adding to, taking from, putting together, taking apart, and comparing, with unknowns in all positions, e.g., by using drawings and equations with a symbol for the unknown number to represent the problem.	Represent and solve problems involving addition and subtraction.
2.OA.2.	Fluently add and subtract within 20 using mental strategies. By end of Grade 2, know from memory all sums of two one-digit numbers.	Add and subtract within 20.
2.OA.3. .	Determine whether a group of objects (up to 20) has an odd or even number of members, e.g., by pairing objects or counting them by 2s; write an equation to express an even number as a sum of two equal addends.	Work with equal groups of objects to gain foundations for multiplication.
2.OA.4	Use addition to find the total number of objects arranged in rectangular arrays with up to 5 rows and up to 5 columns; write an equation to express the total as a sum of equal addends.	
2.NBT.1.	Understand that the three digits of a three-digit number represent amounts of hundreds, tens, and ones; e.g., 706 equals 7 hundreds, 0 tens, and 6 ones. Understand the following as special cases: • 100 can be thought of as a bundle of ten tens — called a "hundred." • The numbers 100, 200, 300, 400, 500, 600, 700, 800, 900 refer to one, two, three, four, five, six, seven, eight, or nine hundreds (and 0 tens and 0 ones).	Understand place value.
2.NBT.2.	Count within 1000; skip-count by 5s, 10s, and 100s.	

SECOND GRADE MATH

Measurement & Data

2.MD.1.	Measure the length of an object by selecting and using appropriate tools such as rulers, yardsticks, meter sticks, and measuring tapes.	Measure and estimate lengths in standard units.
2.MD.2.	Measure the length of an object twice, using length units of different lengths for the two measurements; describe how the two measurements relate to the size of the unit chosen.	
2.MD.3.	Estimate lengths using units of inches, feet, centimeters, and meters.	
2.MD.4.	Measure to determine how much longer one object is than another, expressing the length difference in terms of a standard length unit.	
2.MD.5.	Use addition and subtraction within 100 to solve word problems involving lengths that are given in the same units, e.g., by using drawings (such as drawings of rulers) and equations with a symbol for the unknown number to represent the problem.	Relate addition and subtraction to length.
2.MD.6.	Represent whole numbers as lengths from 0 on a number line diagram with equally spaced points corresponding to the numbers 0, 1, 2, ..., and represent whole-number sums and differences within 100 on a number line diagram.	
2.MD.7.	Tell and write time from analog and digital clocks to the nearest five minutes, using a.m. and p.m.	Work with time and money.
2.MD.8.	Solve word problems involving dollar bills, quarters, dimes, nickels, and pennies, using $ and ¢ symbols appropriately. Example: If you have 2 dimes and 3 pennies, how many cents do you have?	
2.MD.9.	Generate measurement data by measuring lengths of several objects to the nearest whole unit, or by making repeated measurements of the same object. Show the measurements by making a line plot, where the horizontal scale is marked off in whole-number units.	Represent and interpret data.
2.MD.10.	Draw a picture graph and a bar graph (with single-unit scale) to represent a data set with up to four categories. Solve simple put-together, take-apart, and compare problems1 using information presented in a bar graph.	

SECOND GRADE MATH

Geometry

2.G.1.	Recognize and draw shapes having specified attributes, such as a given number of angles or a given number of equal faces.1 Identify triangles, quadrilaterals, pentagons, hexagons, and cubes.	Reason with shapes and their attributes.
2.G.2.	Partition a rectangle into rows and columns of same-size squares and count to find the total number of them.	
2.G.3.	Partition circles and rectangles into two, three, or four equal shares, describe the shares using the words halves, thirds, half of, a third of, etc., and describe the whole as two halves, three thirds, four fourths. Recognize that equal shares of identical wholes need not have the same shape.	

Distance and Speed
1st Grade Activity 3

Step # 1
What is your height in standard units of measure?

Hint: Standard units are those we use most often in America.

Step # 2
Name something 10 times longer.

VOCABULARY FOR GRADE 2
WORD WALL WORDS FOR MATH

Students should illustrate the words for a word wall. Just looking at the cute cards we all like to print and laminate is not particularly effective as a learning tool. While the student created cards take longer (and look significantly messier), the benefits are greater. tThe one who does all the work also does all the learning.

Addition	Value	Yard
Hundreds	Ruler	Meter
Skip count	Yardstick	Number Line
Expanded form	Meterstick	Rows
Standards form	Measuring tape	Halves
Number names	Foot	Thirds

Column	Cents	Angles
Line Plot	Dollar	Trapezoid
Picture Graph	Quarter (coin)	Half-circle
Bar Graph	Faces	Quarter-circle
A.M.	Vertices	Cube
P.M.	Pentagon	Quadrilateral
Hours in a Day / Days in a Week	Octagon	Prism

VOCABULARY FOR GRADE 2
WORD WALL WORDS FOR SCIENCE

DESIGN AND ENGINEERING

construct construction	innovate	robot
data collect, represent, analyze	innovation	rover
design designer	machine	structure
estimate	mechanical	system
foundation	3D printer	weight

INQUIRY	LIFE SCIENCE	
chart	animal	habitat
graph	basic needs	insect
describe	cycle	pet
identify	Earth	plant
observe	forest	pollen
predict		seasons

VOCABULARY FOR GRADE 2
WORD WALL WORDS FOR SCIENCE & ENGINEERING

FORCES & MOTION		SPACE
circular	machine	Earth
circular	objects	moon
fahrenheit	rotation	planet
force	revolution	Sun
gravity	vapor	astronomy

TOOLS		a very few STEM CAREER FIELDS
digital camera	balance **scale**	architect
Internet	digital **scale**	astronomer
magnifing glass	spring **scale**	engineer
3D printer	search engine	medical doctor, nurse, surgeon
ruler	stopwatch	programming
tape measure	telescope	scientist many fields

RHOMBI'S ADVENTURES IN GEOMETRY
SCOPE AND SEQUENCE

	WEEK 1	WEEK 2	WEEK 3	WEEK 4	WEEK 5	WEEK 6	WEEK 7	WEEK 8
RHOMBI'S HOUSE	*********	*********						
WEATHER	ONGOING							
DESIGN	45 MIN	CHALLENGE BRAINSTORM DESIGN BUILD TEST & USE DATA TO MAKE DESIGN CHANGES						
STRUCTURES	60 MIN							
CUBISM	45 MIN							
PERSPECTIVE	30 MIN							
UP ON THE ROOF			*********	*********				
WATER AND SUN			45 MIN	CHALLENGE BRAINSTORM DESIGN BUILD TEST & USE DATA TO MAKE DESIGN CHANGES				
PYRAMID			90 MIN					
ROOF HEIGHT			45 MIN					
MATERIALS			60 MIN					
EROSION			30 MIN					
NEW HOME FOR PET					*********	*********		
HABITATS					90 MIN	CHALLENGE BRAINSTORM DESIGN BUILD TEST & USE DATA TO MAKE DESIGN CHANGES		
ATTRIBUTES					90 MIN			
FOUNDATIONS					45 MIN			
PATTERNS					45 MIN			
SIZE AND SCALE					45 MIN			
PET'S CELEBRATION							*********	*********
DESSERT							45 MIN	CHALLENGE BRAINSTORM DESIGN BUILD TEST & USE DATA TO MAKE DESIGN CHANGES
DIGITAL SCALES							45 MIN	
STRENGTH							60 MIN	
GRAPHICS							90 MIN	
TIMELINES							30 MIN	

Log onto the website to access additional lessons, activities, word wall printables, printable nets, videos, downloadable files, links, and other resources.

www.ten80elementary.com

Please contact us for more information about curriculum, programs, festivals, and professional development.

info@ten80elementary.com

To access the ArtofSTEM website and learn more about fundraising, or to support a school in your area, visit us.

ArtofSTEM.org

CPSIA information can be obtained at www.ICGtesting.com
Printed in the USA
LVOW01*0747081114

412604LV00004B/7/P